Synthesis Lectures on Mechanical Engineering

This series publishes short books in mechanical engineering (ME), the engineering branch that combines engineering, physics and mathematics principles with materials science to design, analyze, manufacture, and maintain mechanical systems. It involves the production and usage of heat and mechanical power for the design, production and operation of machines and tools. This series publishes within all areas of ME and follows the ASME technical division categories.

Shuvra Das

An Introduction to Finite Element Analysis Using Matlab Tools

 Springer

Shuvra Das
Mechanical Engineering Department
University of Detroit Mercy
Detroit, MI, USA

ISSN 2573-3168   ISSN 2573-3176   (electronic)
Synthesis Lectures on Mechanical Engineering
ISBN 978-3-031-17542-8   ISBN 978-3-031-17540-4   (eBook)
https://doi.org/10.1007/978-3-031-17540-4

Preface

The finite element method has become a powerful and ubiquitous tool in the hands of today's engineers. We often come across the term computer-aided engineering (CAE), a methodology used extensively in product development, design and analysis engineering and in numerous other engineering applications. FEA or finite element analysis is a core component of CAE. Many reliable FEA software tools are presently available for engineers to use in solving all classes of engineering problems. User-friendly graphical interfaces have made these tools very easy to use, so much so that engineers can set up, analyze and use the results without a clear understanding of the process. This can lead to an inability to obtain correct results and may lead to flawed decision making. There are numerous textbooks on the subject that do a very good job of explaining all the mathematics behind the method and the fundamental principles of all aspects of the process with detailed mathematical models. The target audience of these textbooks are developers of the FEA method and FEA tools. So, a beginner is often caught between the world of commercial tools that are easy to use and technically dense text books that are hard to understand. This book is an attempt to develop a guide for the user who is interested in learning the method by doing. There is enough discussion of some of the basic theories so that the user can get a broad understanding of the process. And there are many examples with step-by-step instructions for the user to quickly develop some proficiency in using FEA. We have used MATLAB and its PDE toolbox for the examples in this text. The syntax and the modeling process are easy to understand, and a new user can become productive very quickly. The PDE toolbox, just like any other commercial software, can solve certain class of problems well but is not capable of solving every type of problems. For example, it can solve linear problems but is not capable of handling nonlinear problems. Being aware of the capabilities and limitations of any tool is an important lesson for students, and we have, with this book, tried to highlight that lesson as well.

Detroit, USA   Shuvra Das

Contents

About the Author

Shuvra Das started working at University of Detroit Mercy in January 1994 and is currently Professor of Mechanical Engineering. Over this time, he served in a variety of administrative roles such as Mechanical Engineering Department Chair, Associate Dean for Research and Outreach, and Director of International Programs in the College of Engineering and Science. He has an undergraduate degree in Mechanical Engineering from Indian Institute of Technology and Master's and Ph.D. in Engineering Mechanics from Iowa State University. He was a post-doctoral researcher at University of Notre Dame and worked in industry for several years prior to joining Detroit Mercy.

Dr. Das has taught a variety of courses ranging from freshmen to advanced graduate level such as mechanics of materials, introductory and advanced finite element method, engineering design, introduction to mechatronics, mechatronic modeling and simulation, mathematics for engineers, electric drives and electromechanical energy conversion. He led the effort in the college to start several successful programs: an undergraduate major in Robotics and Mechatronic Systems Engineering, a graduate certificate in Advanced Electric Vehicles and thriving partnerships for student exchange with several universities in China.

Dr. Das received many awards for teaching and research at Detroit Mercy as well as from organizations outside the university. His areas of research interest are modeling and simulation of multi-disciplinary engineering problems, engineering education and curriculum reform. He has worked in areas ranging from mechatronics system simulation to multi-physics process simulation using CAE tools such as finite elements and boundary elements. He has authored or co-authored five published books on these topics.

Introduction 1

In today's world of engineering product development modeling and simulation plays a critical role. In an age when computers were not available engineers performed hand calculations, estimations, and a lot of testing to design, manufacture and build products. Many engineering problems are impossible to solve analytically. Over time, engineers and mathematicians devised clever techniques that combined algebraic tricks, graphical techniques, problem similarities from different disciplines, superposition methods, and other ways to simplify calculations to try and solve complicated problems analytically. All these efforts led to many methods such as the use of Thevenin and Norton's theorems in circuit simplifications, Airy stress functions in Elasticity, Energy principles, Castigliano's theorem, area-moment method in drawing shear and bending moment diagrams, and many others. With the advent of computers and development of powerful numerical techniques the use of many of these methods have significantly reduced, rendering them close to being obsolete.

Over time many numerical solution techniques were developed but they could not be properly implemented until computers became available. As computers became more powerful and techniques to solve engineering problems became more and more sophisticated and versatile the role of modeling and simulation became more and more important in product design and development. In fact, as the confidence in numerical techniques grew, the number of physical prototypes needed to "build and test" started to reduce and the product development cycle started to become more efficient, less pricy and fast-moving.

The role of modeling and simulation in product development is often called CAE which is an acronym for Computer Aided Engineering. CAE evolved from a better-known acronym, CAD or computer aided design. The finite element analysis/method (FEA/FEM) is perhaps the best known and most used CAE technique. There are other methods such as the Finite Difference Method and the Boundary Element Method which are sometimes equally good and sometimes even better than FEA in certain applications. But the finite

     1
S. Das, *An Introduction to Finite Element Analysis Using Matlab Tools*,
Synthesis Lectures on Mechanical Engineering,
https://doi.org/10.1007/978-3-031-17540-4_1

element method has become the most popular and universally used method due to its versatility and the availability of the many commercial FEA tools that can be easily used by users without even knowing much about how the tools work.

Modeling and Simulation or computer aided problem solving plays a very critical role in engineering. And we hear of different tools such as commercial FEA tools, Simulink, System modeling tools and others. A very important question that needs to be answered is when and for what kind of problem is a certain tool appropriate? It is this author's experience that students and professionals who are new to this field lack a clear understanding of this issue and hence can get easily frustrated when they are unable to pick the proper tool that would solve their problem satisfactorily. To answer this question, we need to consider the different class of differential equations. A large number of engineering problems can be modeled mathematically using differential equations. Differential equations are broadly divided into two categories: ordinary differential equations and partial differential equations.

Differential equations where primary variables are a function of only one other variable are called ordinary differential equations whereas if the primary variable is a function of more than one variable then the differential equation is a partial differential equation. Therefore, of the following Eqs. 1.1–1.5 the first two are ordinary differential equation while the others are partial differential equations.

$$\frac{d^2X}{dt^2} + kX = F \tag{1.1}$$

$$\frac{d^2T}{dx^2} + hT = hT_0 \tag{1.2}$$

$$\frac{\partial^2T}{\partial x^2} + \frac{\partial^2T}{\partial y^2} + hT = hT_0 \tag{1.3}$$

$$\frac{\partial^2T}{\partial x^2} + \frac{\partial^2T}{\partial y^2} = \frac{\partial T}{\partial t} \tag{1.4}$$

$$\frac{\partial \sigma_x}{\partial x} + \frac{\partial \sigma_{xy}}{\partial y} + f_x = \rho\frac{\partial^2u}{\partial t^2} \quad \frac{\partial \sigma_{xy}}{\partial x} + \frac{\partial \sigma_y}{\partial y} + f_y = \rho\frac{\partial^2v}{\partial t^2} \tag{1.5}$$

All numerical methods for solving these types of equations depend on the process of discretization, i.e., breaking down the domain into small pieces. The domain in any differential equation is either space (denoted by x, y, z etc., the spatial co-ordinates) or time (denoted by t) or both. While the spatial co-ordinates can all be treated similarly and behave similarly, time co-ordinate is somewhat different. Time moves only in one direction, forward. Spatial co-ordinates can be negative or positive and we can define and alter the frame of reference to suit our work. But for time we are usually limited in our freedom. Mostly, for actual problems, time starts now and moves in the positive direction (Fig. 1.1). Also, spatial geometry (Fig. 1.2) has shapes, they have dimensions, surfaces and areas, volumes and boundaries. They have beginning and end points. Time doesn't quite have all of those. This makes problems with spatial variations more work to solve. We treat the time variable a little differently from the spatial variables. This is why even though the first two equations are both ordinary differential equations the first one is dealt with differently than the second one.

Equations where time (e.g., Eq. (1.1)) is the only independent variable are solved using different time integration schemes where the solution at the future time depends on the state at the current time. These classes of problems are very common across engineering. The dependent variable or the primary variable could be any physical quantity such as temperature, voltage, current, velocity, displacement, etc. The equations are called state space equations and tools such as Simulink, Simscape, Dymola, and many others can solve these problems quite efficiently. The most critical feature of these problems that separate them from all the other equations above is the fact that spatial variation of the primary variable does not need to be considered. Hence the spatial domain over which the problem is defined does not need to be broken down or discretized. The primary variable on this spatial domain thus has a single value at any given time. Equation 1.1 in the above list is one such equation.

In the other equations shown, the primary variable is a function of spatial co-ordinates. The variation could be one-, two- or three-dimensional. Spatial co-ordinates have boundaries, usually a finite and known starting point and similar end points. Also, the geometry of the space could be simple or very complex. In most real-world engineering problems the geometry is complex (an engine block, eg.). For complex geometries the differential equations cannot be solved analytically. And the Finite element method is one of the main solution methods of choice for these equations. Equations 1.2 and 1.3 are typical examples where one- and two-dimensional spatial variations are observed. Equation 1.4 is an example where the primary variable is both a function of space as well as time. In these cases the spatial domain needs to be broken up into elements and the solution also has to be calculated at different time intervals. Equation 1.5 is an example where the primary variable is a vector or a tensor and has different components at any particular location. In these cases the resulting equations are a set of coupled partial differential equations.

Fig. 1.1 Time axis

Fig. 1.2 Generic 2-D space

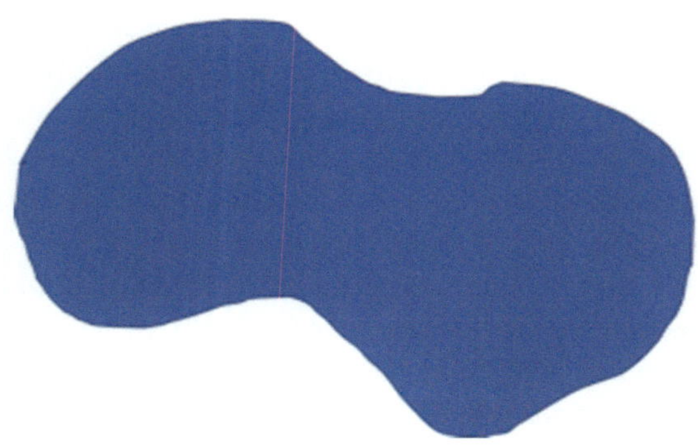

   To provide a very rudimentary understanding of the FEA process let us consider the problem of calculating the circumference of a circle. We know that the circumference can be accurately computed by the formula $2\,\pi R$ where R is the radius of the circle. For the sake of our discussion here let us assume that we try to use an approximate method to find the circumference. Figure 1.3 shows how the circumference may be approximated by dividing it into a number of linear segments. If we take the sum of all the line segments that sum maybe used as an approximate measure of the circumference. As shown in the figure, if we divide the circumference into n linear elements the angle that each element subtends at the center is given by:

$$\theta = \frac{2\pi}{n} \tag{1.6}$$

So half the angle is given by $\frac{\theta}{2} = \frac{\pi}{n}$. Therefore, the length of each element is given by:

$$h_e = 2R\,\mathrm{Sin}(\frac{\theta}{2}) \tag{1.7}$$

Fig. 1.3 FEA process explained through a simple example

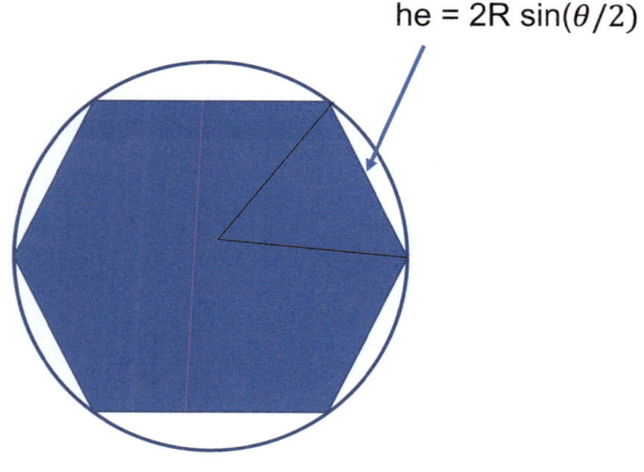

If P_n is an approximate estimate of the circumference, it can be written as:

$$P_n = \sum_{e=1}^{n} h_e = n\left(2R\mathrm{Sin}\left(\frac{\theta}{2}\right)\right) = n\left(2R\mathrm{Sin}\left(\frac{\pi}{n}\right)\right) \tag{1.8}$$

Therefore, this is the approximate estimate of the circle circumference. It is fairly easy to see that as n increases this approximate estimate approaches closer and closer to the true value of the circumference. The concept can also be easily illustrated by computing an estimate of error,

$$\mathrm{Error} = 2\pi R - n\left(2R\mathit{Sin}\left(\frac{\pi}{n}\right)\right) = R\left((2\pi) - 2\pi\frac{\mathit{Sin}\left(\frac{\pi}{n}\right)}{\frac{\pi}{n}}\right) \tag{1.9}$$

The error approaches zero the higher n becomes.

This was a simple illustration of the FEA method and how it can be used to compute approximate solutions of real problems. It will be important to point out that approximate doesn't mean inaccurate. Accuracy could be easily increased by increasing the mesh or discretization level.

1.1   FEA Process

The finite element solution process has sound mathematical basis and we will be getting into more of the necessary details of that but it is a good idea to have a big-picture understanding of the process itself. Since analytical solution of the differential equation is not possible for the whole domain, the domain is divided into smaller parts, known as elements. The solution for the primary variables is not known but its behavior is approximated using known functions (linear, quadratic, etc.) over each of these elements using the unknown values at nodes (element corners or endpoints). The elements are then assembled mathematically for the entire geometry and this assembly results in a set of linear equations where the unknowns are nodal values of the primary variables. Upon solving this set of linear equations, we can compute the primary variable at discrete points all across the geometry. The solution at all other locations on the geometry is then obtained by interpolating the nodal values using the same approximation that was initially assumed over each element. Accuracy is improved by reducing element size (or increasing the number of elements) and/or by choosing better approximation schemes over each element (also known as choosing higher order elements).

The flowchart (Fig. 1.4) provides a broad description of the FEA process for a typical problem. This is essentially the standardized process path and all commercially available tools implement this in some form. In the next few paragraphs, we have summarized the different steps in the flowchart.

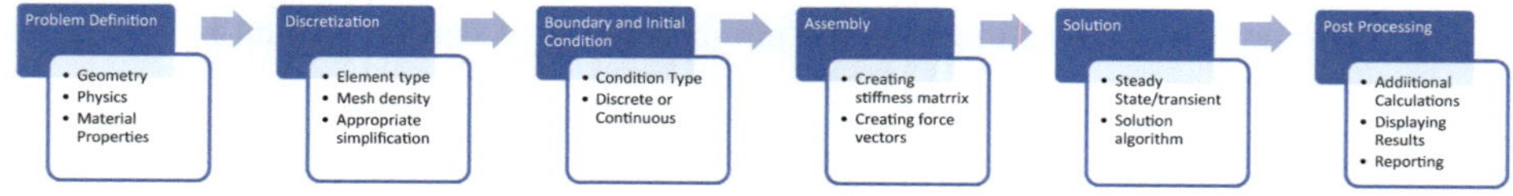

Fig. 1.4 The FEA process

1.1.1 Problem Definition

At this stage of the process, one needs to define the geometry/domain, boundary and the problem type such as stress analysis, heat transfer, fluid flow, combined physics, etc. Once the problem physics and geometry are defined, appropriate material properties need to be specified. For example, for elastic stress analysis, Young's Modulus and Poisson's ratio need to be specified, if it is a steady state thermal problem then the thermal conductivity needs to be provided or if it is a transient thermal problem then thermal conductivity, density and specific heat needs to be provided. Figure 1.5 shows a typical picture at this stage of the process.

1.1.2 Discretization

At this stage the problem geometry needs to be broken down into smaller regions or elements (Fig. 1.6). The elements are designated as linear, quadratic, cubic, etc. depending on the assumed variation of the primary variable on the element. A key aspect of this step is mesh density, i.e. how coarse or fine the mesh will be. Solution accuracy and speed of computation both depends on the mesh density. The larger the number of elements the more fine the mesh is and the more accurate is the result—this statement is

Fig. 1.5 Geometry and problem definition

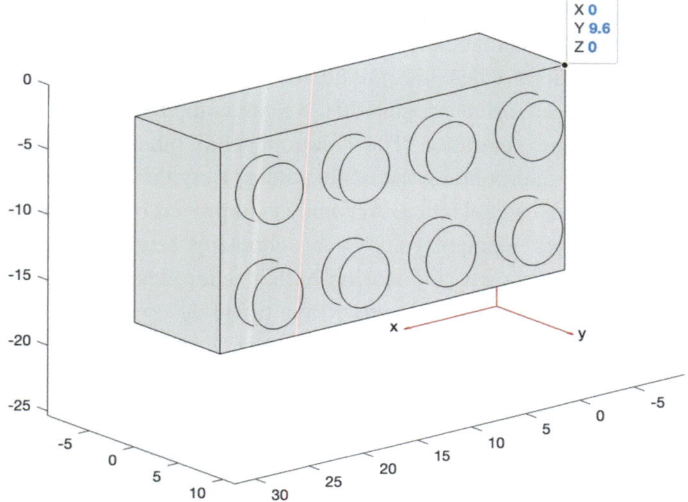

Fig. 1.6 Meshing or discretization

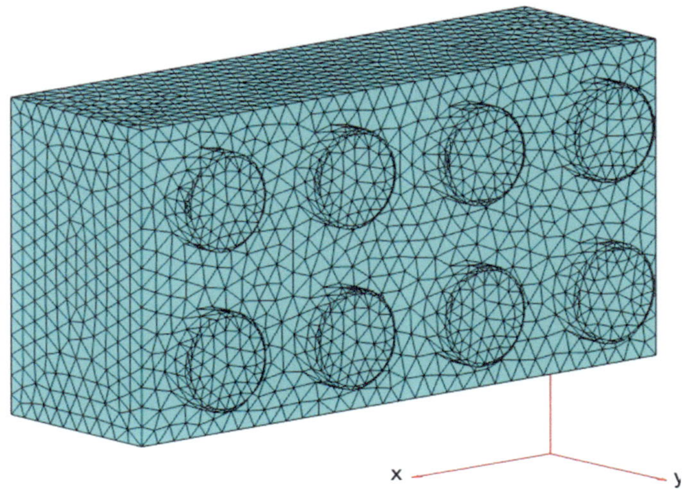

generally true most of the time. Also, the larger the size of the problem the longer it takes to solve. Until a few years ago computational speed and cost of computation were issues that were very important so analysts had to strike a balance between accuracy and computation cost. Given the rate at which computational power has increased this issue has become less important.

1.1.3   Boundary and Initial Conditions

Solution of any problem is dictated by the boundary conditions. Two problems with identical geometry and properties can have different solutions if the boundary conditions are different. At this stage of the process the correct boundary conditions have to be properly applied so that the resulting solution is correct. Also, if the problem is transient or time dependent, the initial conditions need to be defined as well. Figure 1.7 shows a problem with graphical indications showing how boundary conditions are applied.

1.1.4   Assembly

When all the conditions that properly defines a problem are clearly accounted for the resulting equations have to be assembled into matrix–vector form. The physical manifestation of assembly may be thought of as gathering up or assembling all the small elements to make the problem whole again. The matrix–vector equations are then ready for solution.

Fig. 1.7 Applying boundary conditions

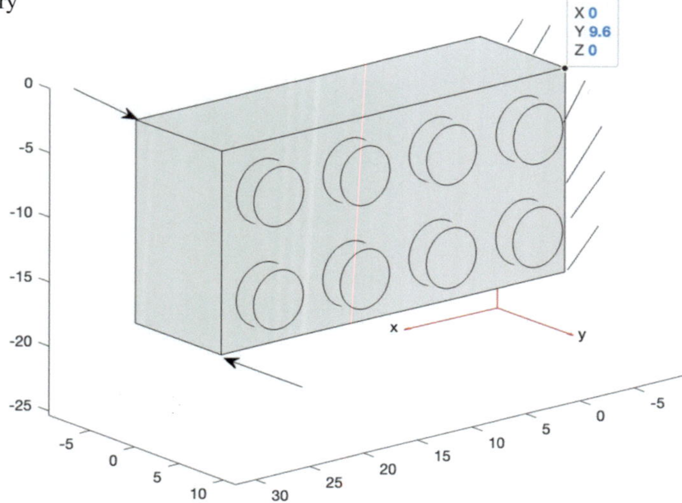

Fig. 1.8 Displacement visualization

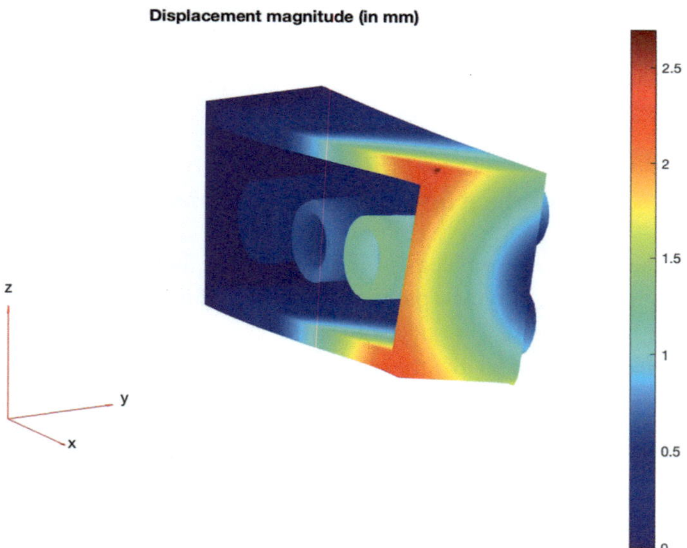

1.1.5   Solution

The large equation sets are solved for the primary variable (displacement for stress analysis problems, temperature for thermal analysis problems, etc.). There are many well-known and efficient algorithms that are used for solving large sets of linear equations.

1.1.6   Post Processing

Post processing is a critical step in the Finite element based problem solution process. Once the primary variables are calculated they can be used to calculate a variety of secondary variables (for example different types of stresses in case of structural problems or heat flux in case of heat transfer problems) (Fig. 1.8). The other key task at this stage of the process is displaying results, usually in graphical forms so that the user and others can easily use the results to draw conclusions, understand and use the problem solution to make appropriate design decisions.

1.2   Some Relevant Matlab Information

In this text we are introducing some basic concepts of Finite Element Method (FEA) and Matlab as a tool for performing FEA analyses. There are two ways to use Matlab for FEA work. The core calculations in FEA centers around Matrix operations. Matlab is very well suited for performing these types of calculations; Matlab's origin is from Matrix operations. For simpler problems users can use Matlab commands to perform these analyses very easily or develop and use Matlab codes to automate the process of FEA problem setup and solution. Matlab also has a PDE Toolbox (partial differential equation toolbox). This toolbox is used to solve many types of linear physical problems using the FEA method. Particularly, this toolbox is very useful for solving linear Finite Element problems with applications in domains such as structural mechanics, heat transfer and electromagnetics. The PDE toolbox is capable of performing the various steps that were identified in this chapter, such as geometry and meshing, boundary condition application, solving a variety of equation types, and post processing. In this text we have included examples of both type of work. For some of the simpler and earlier examples we use basic Matlab command to develop code for solving certain problems. And for problems that are a little more involved we have demonstrated the use of the PDE toolbox.

In the following sections we will review some basic matrix operations that are core features for any FEA work and introduce the PDE toolbox. Like with every other Matlab module, there is extensive documentation on the PDE toolbox available online for users to refer to. We are not trying to be comprehensive or exhaustive in this introduction. So users will benefit a lot by using Matlab resources.

1.2.1   Basic Matrix Operations Using MATLAB

Among many other capabilities MATLAB can be used to perform all Matrix operations. Here is a short summary of some of them.

Matrix Multiplication, transpose, inversion, and solution of linear equations.

```
A = [1 5 9;-2 3 7;3 3 11]

A = 3×3
     1      5      9
    -2      3      7
     3      3     11
```

```
B = [1 7 1;2 5 1;3 -3 1]

B = 3×3
     1      7      1
     2      5      1
     3     -3      1
```

```
C = A*B

C = 3×3
    38      5     15
    25    -20      8
    42      3     17
```

```
D = B*A

D = 3×3
   -10     29     69
    -5     28     64
    12      9     17
```

Transpose of a Matrix

```
transpose(A)

ans = 3×3
     1     -2      3
     5      3      3
     9      7     11
```

Determinant

```
det(B)

ans = -6
```

```
det(A)

ans = 92
```

Inverse

```
inv(A)

ans = 3×3
    0.1304    -0.3043     0.0870
    0.4674    -0.1739    -0.2717
   -0.1630     0.1304     0.1413
```

Solving a system of Linear equation

```
[A] {X} = {b}
{X} = inv([A]) {b}


A

  A = 3×3
        1      5      9
       -2      3      7
        3      3     11


b = [3;7;9]

  b = 3×1
        3
        7
        9


X = inv(A)*b; %solving for the unknowns using matrix inverse

  X = 3×1
      -0.9565
      -2.2609
       1.6957


X = A\b; %solving for the unknowns using more efficient built-in algorithms

  X = 3×1
      -0.9565
      -2.2609
       1.6957
```

1.2.2   PDE Toolbox

Partial Differential Equation Toolbox™ is capable of solving structural mechanics, heat transfer, and general partial differential equations (PDEs) using finite element analysis.

One can perform linear static analysis of structural problems to compute deformation, stress and strain. It can perform dynamic analysis as well through direct time integration and modal superposition. Natural frequency and mode shape computation for structures is also possible. The toolbox can perform heat transfer problems where the dominant mode is conduction. Convective and Radiative boundary conditions are easily handled as well. Certain multi-physics problems such as thermal stress analysis where transient heat transfer data is used to compute the stress can be solved. You can also solve standard problems such as diffusion, electrostatics, and magnetostatics, as well as custom PDEs.

Partial Differential Equation Toolbox allows the import of 2D and 3D geometries from STL or mesh data. Mesh generation is automated with triangular and tetrahedral elements. Extensive postprocessing capability is also available within the toolbox.

It is important to also understand the limitations of the toolbox. This tool is not suitable to perform non-linear problems, such as material non-linearities involved in plastic deformation or non-linear behavior due to problem physics in certain fluid flow problems. In general, fluid flow problems and related heat transfer problems where convection is dominating the process is not for this tool. There are many commercial tools in the marketplace for solving those types of problems. Despite these shortcomings, the PDE toolbox is still quite powerful and is a tool that can be quickly mastered to solve an extensive class of problems.

1.3 Summary

In this introductory chapter we have provided a very broad overview of the FEA process and how we pick what tool to use for solving different types of engineering problems. In the next few chapters we will explore the process in more details. The next few chapters cover the following topics:

Chapter 2, Discrete FEA.

Chapter 3, Continuum FEA and its use in Heat Transfer problems.

Chapter 4, Continuum FEA and its use in Stress Analysis problems.

Chapter 5, Solving more complex problems such as thermal stress analysis and electromagnetics.

Discrete Finite Elements

Early applications of the finite element method came from the world of structural problems where large, discrete structures such as trusses had to be analyzed. Trusses or or other structures made of discrete members that are only connected to each other at their ends are naturally conducive to the finite element process. The force displacement relationships for each member can be written out separately and then assembled to form a matrix-vector system of equations for the whole structure. This method was originally called matrix method of structural analysis and is currently referred to as discrete finite element method. In this chapter we will explore the discrete FEA method basics and how it is implemented for static and dynamic problems in structural mechanics.

2.1 Static Problems

We start this discussion with the definition of a basic 2-D spring or strut element. This is the building block for all discrete FEA problems. Consider a strut, a long and thin rod, that is a single element in an elastic structure. It can be treated as an elastic member or a spring. From basic mechanics of materials, we can write the force displacement equation as:

$$P = \frac{AE}{L}\Delta \tag{2.1}$$

where Δ is the displacement, P is the force, A is the cross-sectional area of the member/element, L is its length, and E is the modulus of elasticity (Fig. 2.1).

This equation may be written as a spring equation by relating the force in the member (spring) to its displacement. Therefore, the ratio that combines the geometric and material parameters, A, E, and L can be represented as the spring constant.

© The Author(s), under exclusive license to Springer Nature Switzerland AG 2023   13
S. Das, *An Introduction to Finite Element Analysis Using Matlab Tools*,
Synthesis Lectures on Mechanical Engineering,
https://doi.org/10.1007/978-3-031-17540-4_2

Fig. 2.1 Schematic for an elastic spring element

So the above equation can be written as:

$$P = \frac{AE}{L}\Delta = k\Delta \tag{2.2}$$

If we write this simple force-displacement relationship for a general element whose end points are called locations 1 and 2 and the forces at the two end points and the corresponding displacements are represented as f1x and f2x and d1x and d2x respectively, then the above relationship may be extended as:

$$f_{2x} = -f_{1x} \tag{2.3}$$

$$f_{2x} = k(d_{2x} - d_{1x}) = k(-d_{1x} + d_{2x}) \tag{2.4}$$

$$f_{1x} = -k(d_{2x} - d_{1x}) = k(-d_{2x} + d_{1x}) = k(d_{1x} - d_{2x}) \tag{2.5}$$

Writing Eqs. 2.4 and 2.5 together in matrix form we get:

$$\begin{bmatrix} f_{1x} \\ f_{2x} \end{bmatrix} = \begin{bmatrix} k & -k \\ -k & k \end{bmatrix} \begin{bmatrix} d_{1x} \\ d_{2x} \end{bmatrix} \tag{2.6}$$

$$[f] = [K][d] \tag{2.7}$$

The vector on the left is known as the force vector, the matrix on the right-hand side is the stiffness matrix and the vector that is multiplied with it is the displacement vector.

This is the force-displacement equation for a basic elastic element. Note that matrix equation is a set of two equations with four unknowns, namely the two forces at the two nodes and the two displacements at the same two nodes. The stiffness matrix is known since k is made of geometric and material parameters. In order to solve such an equation at least two of the four unknowns need to be specified. This knowledge/information will be obtained from boundary conditions, which are indicative of this element's interaction with the world outside. But, more on that will be discussed a little later.

Example 2.1 Consider a structure that is made of two such elements (AB and BC):

The structure is made of two different sections shown in the figure as AB and BC. There is a force P acting at A as shown in the figure and the end C is clamped or fixed permanently.

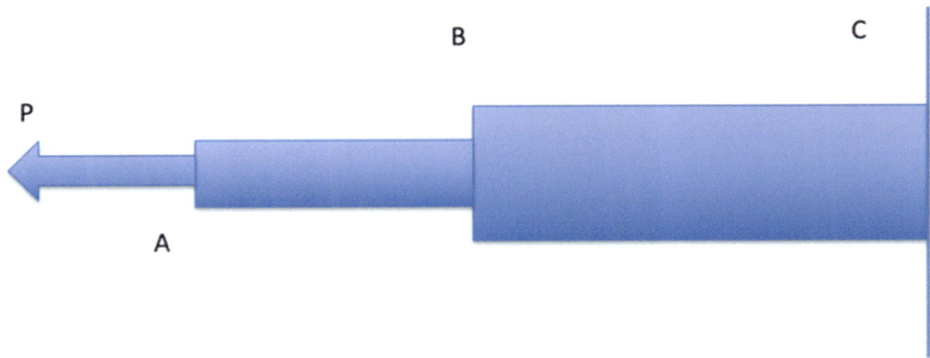

Fig. 2.2 A loaded structure

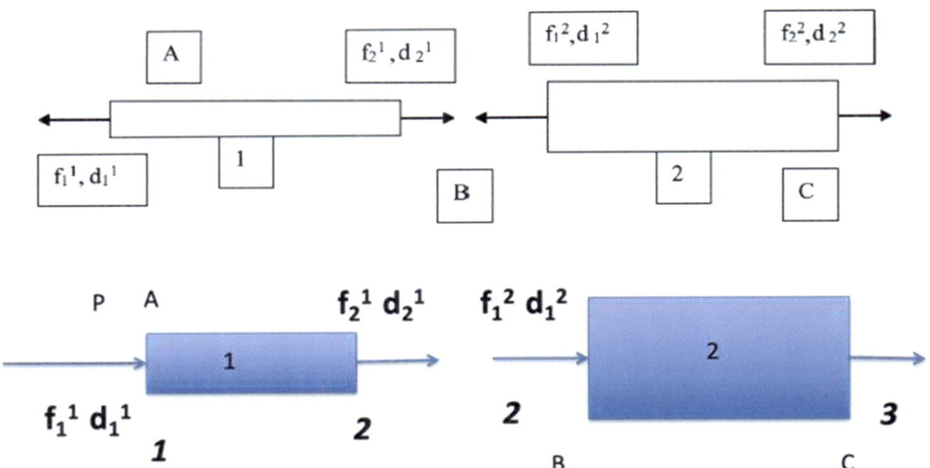

Fig. 2.3 Parts of the structure in Fig. 2.2

Figure 2.2 for the example is redrawn in Fig. 2.3 after separating the two pieces into separate pieces and they are then called element 1 and element 2. Both elements have two endpoints, namely node 1 and node 2 (numbered from left to right). Associated with this concept is the global and the local node numbering system. The global node numbers of the three nodes are 1, 2, and 3 (numbered from left to right). Whereas in each element there is a node 1 and node 2; these are local node numbers. The figure also shows the forces and displacements at each node for each element. The notation used here has a subscript and a superscript associated with it. The subscript is for the local node number and the superscript is for the element number.

Therefore, using the basic elemental equation derived earlier the equations for both elements may be written as:

For element #1

$$\begin{bmatrix} f_1^1 \\ f_2^1 \end{bmatrix} = \begin{bmatrix} k & -k \\ -k & k \end{bmatrix} \begin{bmatrix} d_1^1 \\ d_2^1 \end{bmatrix} \tag{2.8}$$

For element #2

$$\begin{bmatrix} f_1^2 \\ f_2^2 \end{bmatrix} = \begin{bmatrix} k & -k \\ -k & k \end{bmatrix} \begin{bmatrix} d_1^2 \\ d_2^2 \end{bmatrix} \tag{2.9}$$

Now, using some basic MATLAB commands we can create these matrices.

```
% Definition of Stiffness Matrices
k = [100,200]

k = 1×2
    100    200

kel1 = 100*[1 -1;-1 1]

kel1 = 2×2
    100   -100
   -100    100

kel2 = 200*[1 -1;-1 1]

kel2 = 2×2
    200   -200
   -200    200

% Initializing
elem_con = [1 2; 2 3]; %element connectivity

elem = size(elem_con,1);%total nummber of elements
nodes = elem+1;% in a 1 D system #nodes = #elements+1
K = zeros(nodes);%Initialize global stiffness matrix
```

We will now need to put these two sets of equations together (assemble) so that we can get one set of equations for the whole system. In order to do that we will have to consider what happens at the junction. The second node of the first element is the same as the first node of the second element and in reality, these two points are one and the same. This means that magnitude of the displacement at this point is unique. Mathematically, therefore,

$$d_2^1 = d_1^2 \tag{2.10}$$

Also, when these two nodes are joined the forces shown at this node on each of the elements sum up to be equal to the external force acting at this point. Mathematically,

$$f_2^1 + f_1^2 = F_2 \tag{2.11}$$

where F2 is the net external force at global node 2 (which happens to be zero in this case).

So, if we try to put the two sets of equations together by adding the second equation for the first element and the first equation for the second element we get:

$$f_2^1 + f_1^2 = -k_1 d_1^1 + (k_1 + k_2)d_2^1 - k_2 d_2^2 \tag{2.12}$$

The other two equations remain intact. Now putting all of them together we get:

$$\begin{bmatrix} f_1^1 \\ f_2^1 + f_1^2 \\ f_2^2 \end{bmatrix} = \begin{bmatrix} k_1 & -k_1 & 0 \\ -k_1 & k_1 + k_2 & -k_2 \\ 0 & -k_2 & k_2 \end{bmatrix} \begin{bmatrix} d_1^1 \\ d_2^1 \\ d_2^2 \end{bmatrix} \tag{2.13}$$

$$\begin{bmatrix} F_1 \\ F_2 \\ F_3 \end{bmatrix} = \begin{bmatrix} k_1 & -k_1 & 0 \\ -k_1 & (k_1 + k_2) & -k_2 \\ 0 & -k_2 & k_2 \end{bmatrix} \begin{bmatrix} d_1 \\ d_2 \\ d_3 \end{bmatrix} \tag{2.14}$$

where the Fs represent the net forces at the nodes and the ds represent the displacements at the nodes.

These three Eqs. (2.14) represent the complete set of force displacement equations for the two-element structure in this example. These three equations are known as the assembled equations for the whole system. For a system that is small enough (in this case it has two elements the assembled stiffness matrix can also be replaced with actual values for k1 and k2. So, K the stiffness matrix can be written as:

$$[K] = \begin{bmatrix} k_1 & -k_1 & 0 \\ -k_1 & k_1 + k_2 & -k_2 \\ 0 & -k_2 & k_2 \end{bmatrix} = \begin{bmatrix} 100 & -100 & 0 \\ -100 & 300 & -200 \\ 0 & -200 & 200 \end{bmatrix} \tag{2.15}$$

Assembly of a stiffness matrix in FEA is a crucial step in the solution process. While doing this by hand is easy for smaller number of elements, it becomes tedious for larger problems and close to impossible for multi-dimensional problems. The process is usually automated in any FEA solution process. The following few lines of code automates it for a 1-D spring element. This will work for any number of elements.

```
%Assembly
for i=1:elem
    ke = k(i)*[1 -1;-1 1]
    node1 = elem_con(i,1);
    node2 = elem_con(i,2);
    K(node1,node1) = K(node1,node1) + ke(1,1);
    K(node1,node2) = K(node1,node2) + ke(1,2);
    K(node2,node1) = K(node2,node1) + ke(2,1);
    K(node2,node2) = K(node2,node2) + ke(2,2);
end

ke = 2×2
     100   -100
    -100    100
ke = 2×2
     200   -200
    -200    200


K


K = 3×3
     100   -100      0
    -100    300   -200
       0   -200    200
```

Note once again that the three equations have a total of six unknowns, the three forces and the three displacements. In order to use these equations to solve for some of the unknown quantities it is necessary to provide information about three of the six unknowns. This info is obtained from the boundary conditions. Another way of considering the boundary conditions is to consider all the unknowns and determining which one we know and which one we don't. Let's look at the forces first. F1 is the force at the node 1 and it is shown to be acting in the negative x direction and is of magnitude P. So,

$$F_1 = -P \tag{2.16}$$

F2 is the external force at node 2 and is equal to zero since there are no forces at that node.

$$F_2 = 0 \tag{2.17}$$

F3 is the reaction force at the fixed end. At this point we don't know what it is (actually from basic mechanics we do know it but we will assume that it is not known).

$$F_3 = ? \tag{2.18}$$

Of the three displacements d1 and d2 are not known but d3 is zero since that end is fixed or clamped. So,

$$d_1 = ?$$
$$d_2 = ?$$
$$d_3 = 0 \tag{2.19}$$

Hence of the six unknowns we now know three (F1, F2 and d3) and the other three (F3, d1 and d2) needs to be determined. So, the system of equations may be written as:

$$\begin{bmatrix} -P \\ 0 \\ F_3 \end{bmatrix} = \begin{bmatrix} k_1 & -k_1 & 0 \\ -k_1 & (k_1+k_2) & -k_2 \\ & -k_2 & k_2 \end{bmatrix} \begin{bmatrix} d_1 \\ d_2 \\ 0 \end{bmatrix} \tag{2.20}$$

```
P = -15;
%Forces
F = [P;0;0]

F = 3×1
    -15
      0
      0


d = zeros(nodes,1)

d = 3×1
     0
     0
     0

d1 = 0;
```

To solve these equations, we can separate out the first two equations and solve them separately for d1 and d2 and then solve the third equation for F3. After separating the first two equations and reducing it by throwing away the zeros we get:

$$\begin{bmatrix} -P \\ 0 \end{bmatrix} = \begin{bmatrix} k_1 & -k_1 \\ -k_1 & k_1+k_2 \end{bmatrix} \begin{bmatrix} d_1 \\ d_2 \end{bmatrix} \tag{2.21}$$

Inverting the stiffness matrix and multiplying the left-hand side with the inverted matrix will provide the solution for the displacements.

$$[K]^{-1}[F] = [d] \tag{2.22}$$

$$F_3 = -k_2 d_1 + k_2 d_2 \tag{2.23}$$

```
% Solve the system of equations for the unknown displacements
d(1:2) = K(1:2,1:2)\F(1:2)

  d = 3×1
     -0.2250
     -0.0750
           0
```

```
% And use the displacement vector to calculate all the unknown forces
% Such as the reaction forces.
F = K*d

  F = 3×1
     -15
       0
      15
```

```
% Use the displacement of each element or the end diplacement of the
two
% nodes to calculate the forces in each element
del1 = d(1:2)

  del1 = 2×1
     -0.2250
     -0.0750
```

```
del2 = d(2:3)

  del2 = 2×1
     -0.0750
           0
```

```
fel1 = kel1*del1

  fel1 = 2×1
     -15
      15
```

```
fel2 = kel2*del2

  fel2 = 2×1
     -15
      15
```

And if we want to find the stresses in each element, we can find that by using the displacement to strain to stress calculation.

$$\sigma = E\epsilon = E\left(\frac{d_2 - d_1}{L}\right) \tag{2.24}$$

where E is the Young's modulus and L is the length of the element.

Example 2.2 Consider a case of one additional element

If the number of elements increase the size of the stiffness matrix increases as well. So for three elements the stiffness matrix will be:

$$\begin{bmatrix} k_1 & -k_1 & 0 & 0 \\ -k_1 & k_1 + k_2 & -k_2 & 0 \\ 0 & -k_2 & k_2 + k_3 & -k_3 \\ 0 & 0 & -k_3 & k_3 \end{bmatrix} \tag{2.25}$$

```
clear % clear the memory of all values calculated prior to the start
of this problem
k(1) =100%The slider in the code allows for altering the force

k = 100

k(2) =200%The slider in the code allows for altering the force

k = 1×2
     100    200

k(3) =400%The slider in the code allows for altering the force

k = 1×3
     100    200    400


% Definition of Stiffness Matrices

kel1 = k(1)*[1 -1;-1 1];
kel2 = k(2)*[1 -1;-1 1];
kel3 = k(3)*[1 -1;-1 1];
```

```matlab
% Initializing the systemm
elem_con = [1 2; 2 3;3 4] %elemment connectivity
```

```
elem_con = 3×2
       1       2
       2       3
       3       4
```

```matlab
elem = size(elem_con,1)%total number of elements
```

```
elem = 3
```

```matlab
nodes = elem+1 %in a 1 D system #nodes = #elements+1
```

```
nodes = 4
```

```matlab
K = zeros(nodes)%Initialize global stiffness matrix
```

```
K = 4×4
       0       0       0       0
       0       0       0       0
       0       0       0       0
       0       0       0       0
```

```matlab
%Assembly
for i=1:elem
    ke = k(i)*[1 -1;-1 1]
    node1 = elem_con(i,1);
    node2 = elem_con(i,2);
    K(node1,node1) = K(node1,node1) + ke(1,1);
    K(node1,node2) = K(node1,node2) + ke(1,2);
    K(node2,node1) = K(node2,node1) + ke(2,1);
    K(node2,node2) = K(node2,node2) + ke(2,2);
end
```

```
ke = 2×2
     100    -100
    -100     100
ke = 2×2
     200    -200
    -200     200
ke = 2×2
     400    -400
    -400     400
```

```matlab
K
```

```
K = 4×4
```

```
    100   -100      0      0
   -100    300   -200      0
      0   -200    600   -400
      0      0   -400    400
```

```matlab
% Apply the External Load at a certain node.  The slider in the code allows
% for altering the force value
P = 70
```

```
 P = 70
```

```matlab
%Forces
F = [P;0;0;0]
```

```
 F = 4×1
      70
       0
       0
       0
```

```matlab
d = zeros(nodes,1)
```

```
 d = 4×1
       0
       0
       0
       0
```

```matlab
d1 = 0;
d(1:3) = K(1:3,1:3)\F(1:3) % computing the nodal displacements
```

```
 d = 4×1
     1.2250
     0.5250
     0.1750
          0
```

```matlab
F = K*d
```

```
 F = 4×1
    70.0000
    -0.0000
     0.0000
   -70.0000
```

```matlab
del1 = d(1:2);
```

Figure 2.4 shows the displacements across the length of the problem.

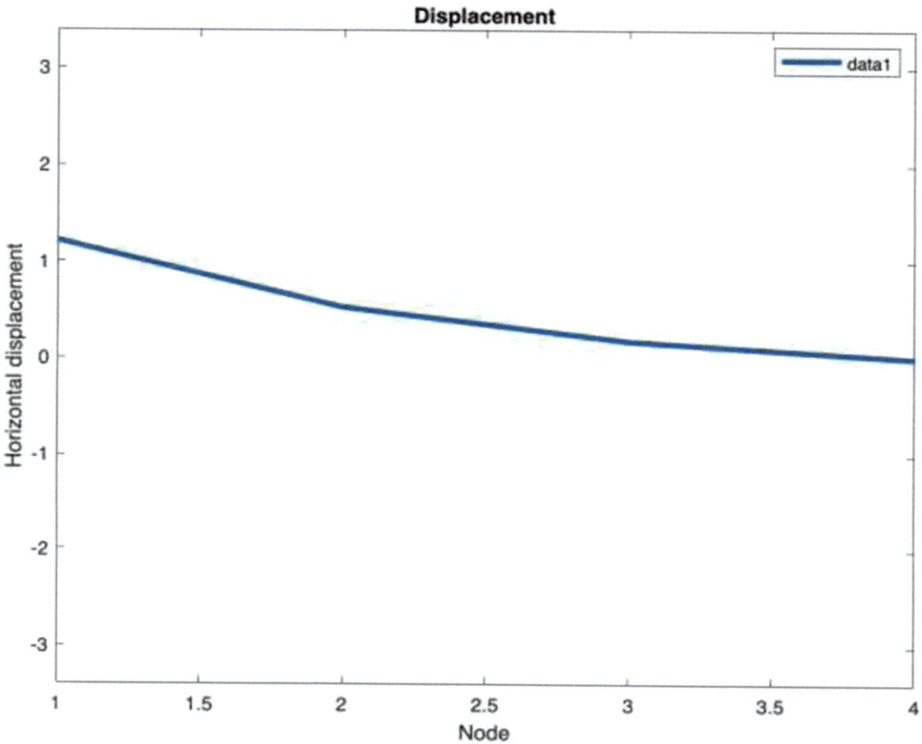

Fig. 2.4 Displacement across the domain in Example 2.2

```
del2 = d(2:3);
del3 = d(3:4);
fel1 = kel1*del1;
fel2 = kel2*del2;
fel3 = kel3*del3;
plot(d,'LineWidth',3); % plotting the displacements at the nodes

%Plot properties
xlim([1.00 4.00])
ylim([-3.40 3.40])
%legend('show')
title('Displacement ')
xlabel('Node')
ylabel('Horizontal displacement')
```

Example 2.3 In this example we have 4 elements shown as springs. Each spring has a spring constant of 200 N/mm. The left end (node 1 is fixed). There is a force of 100 N at node 2 and the total displacement at node 5 is 5 mm. Set up the equations and solve for the unknown force at the 5th node and the displacement at all the nodes (Fig. 2.5).

$$
\begin{bmatrix} F1 \\ F2 \\ F3 \\ F4 \\ F5 \end{bmatrix} = \begin{bmatrix} k1 & -k1 & 0 & 0 & 0 \\ -k1 & k1+k2 & -k2 & 0 & 0 \\ 0 & -k2 & k2+k3 & -k3 & 0 \\ 0 & 0 & -k3 & k3+k4 & -k4 \\ 0 & 0 & 0 & -k4 & k4 \end{bmatrix} \begin{bmatrix} d1 \\ d2 \\ d3 \\ d4 \\ d5 \end{bmatrix} \tag{2.26}
$$

After applying the given conditions:

$$
\begin{bmatrix} F1 \\ 100 \\ 0 \\ 0 \\ F5 \end{bmatrix} = \begin{bmatrix} k1 & -k1 & 0 & 0 & 0 \\ -k1 & k1+k2 & -k2 & 0 & 0 \\ 0 & -k2 & k2+k3 & -k3 & 0 \\ 0 & 0 & -k3 & k3+k4 & -k4 \\ 0 & 0 & 0 & -k4 & k4 \end{bmatrix} \begin{bmatrix} 0 \\ d2 \\ d3 \\ d4 \\ 5 \end{bmatrix} \tag{2.27}
$$

The strategy to solve these equations is to solve the three middle equations for d2, d3 and d4 and then solve the first and the last equation for F1 and F5. But since d5 is non zero we will have to make sure that for the fourth equation -d5*k4 is properly accounted for.

```
clear % clear the memory of all values calculated prior to the start
of this example
k(1)=200
```

```
k = 200
```

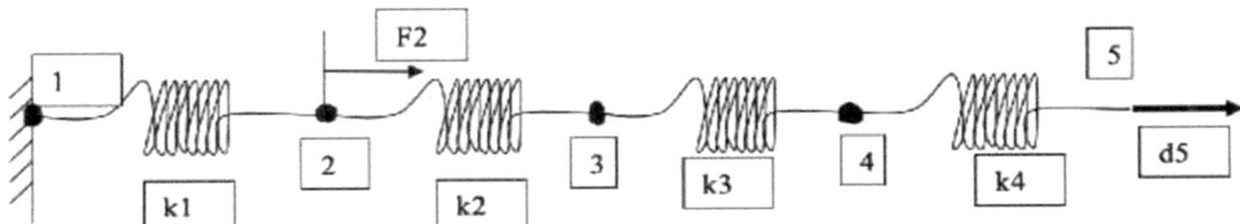

Fig. 2.5 Problem definition for Example 2.3

```
k(2)=200

k = 1×2
   200    200


k(3)=200

k = 1×3
   200    200    200


k(4)=200

k = 1×4
   200    200    200    200

% Definition of Stiffness Matrices

kel1 = k(1)*[1 -1;-1 1];
kel2 = k(2)*[1 -1;-1 1];
kel3 = k(3)*[1 -1;-1 1];
kel4 = k(4)*[1 -1;-1 1];



% Initializing the system
elem_con = [1 2; 2 3;3 4;4 5] %elemment connectivity

elem_con = 4×2
     1     2
     2     3
     3     4
     4     5


elem = size(elem_con,1)%total nnummber of elements

elem = 4

nodes = elem+1 %in a 1 D system #nodes = #elements+1

nodes = 5

K = zeros(nodes)%Initialize global stiffness matrix

K = 5×5
     0     0     0     0     0
     0     0     0     0     0
     0     0     0     0     0
     0     0     0     0     0
```

```
     0        0        0        0        0
```

```matlab
%Assembly
for i=1:elem
    ke = k(i)*[1 -1;-1 1]
    node1 = elem_con(i,1);
    node2 = elem_con(i,2);
    K(node1,node1) = K(node1,node1) + ke(1,1);
    K(node1,node2) = K(node1,node2) + ke(1,2);
    K(node2,node1) = K(node2,node1) + ke(2,1);
    K(node2,node2) = K(node2,node2) + ke(2,2);
end
```

```
ke = 2×2
    200   -200
   -200    200
ke = 2×2
    200   -200
   -200    200
ke = 2×2
    200   -200
   -200    200
ke = 2×2
    200   -200
   -200    200
```

```matlab
K
```

```
K = 5×5
    200   -200      0      0      0
   -200    400   -200      0      0
      0   -200    400   -200      0
      0      0   -200    400   -200
      0      0      0   -200    200
```

```matlab
P = 100
```

```
P = 100
```

```matlab
%Forces
F = [0;P;0;0;0]
```

```
F = 5×1
     0
   100
```

```
          0
          0
          0
```

```
d = zeros(nodes,1)
```

```
d = 5×1
     0
     0
     0
     0
     0
```

```
d1 = 0;
d5 = 5;
d(5,1) = d5
```

```
d = 5×1
     0
     0
     0
     0
     5
```

```
F=F-K*d
```

```
F = 5×1
         0
       100
         0
      1000
     -1000
```

```
% Solving for displacement of nodes 2, 3, 4
d(2:4) = K(2:4,2:4)\F(2:4)
```

```
d = 5×1
         0
    1.6250
    2.7500
    3.8750
    5.0000
```

```
F = K*d
```

```
F = 5×1
  -325.0000
   100.0000
    -0.0000
    -0.0000
   225.0000

plot(d,'LineWidth',3); % plotting the displacement at the nodes

%Plot properties
%legend('show')
title('Displacement ')
xlabel('Node')
ylabel('Horizontal displacement')


plot(F,'LineWidth',3); %plotting the forces at each node

%Plot properties
%legend('show')

title('Force ')
xlabel('Node')
ylabel('Force')
```

Figure 2.6 shows the horizontal displacements at the nodes and Fig. 2.7 shows the forces at each of the nodes.

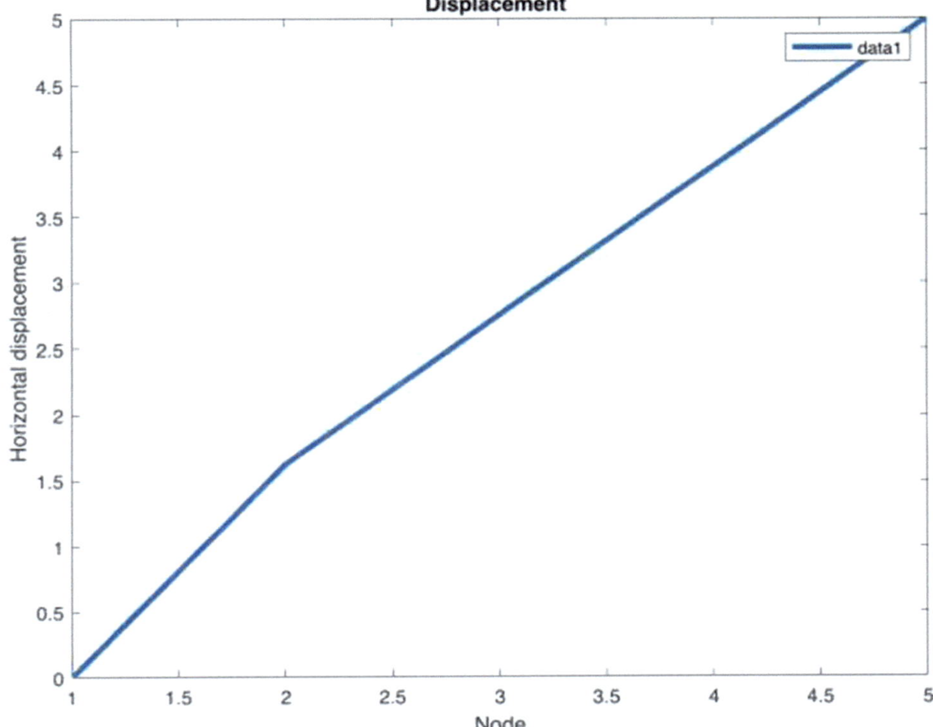

Fig. 2.6 Horizontal displacement across the domain

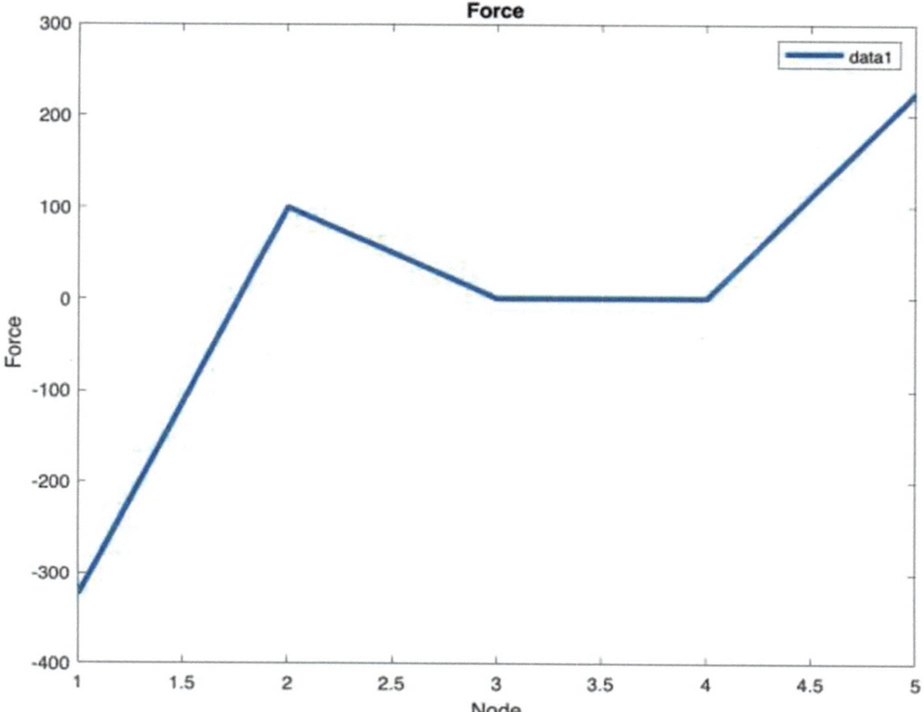

Fig. 2.7 Forces across the domain

Example 2.4 In order to explore the process of stress calculation in elastic elements we take a look at Example 2.1 with some changes. Figure 2.8 shows the same figure as before. Let the length of AB be 1 m and length BC be 1.5 m. Let the area of cross section for AB be 200 mm^2 and area of cross-section for BC be 400 mm^2 and the Youngs moduli be 200 GPa for AB and 150 GPa for BC. Let the force P shown in the figure be 50 KN. We will redo this problem with this new information and this time we will also compute the stresses in each region AB and BC.

Now, using some basic MATLAB commands we can create these matrices and solve for displacements at the nodes and stresses in the elements.

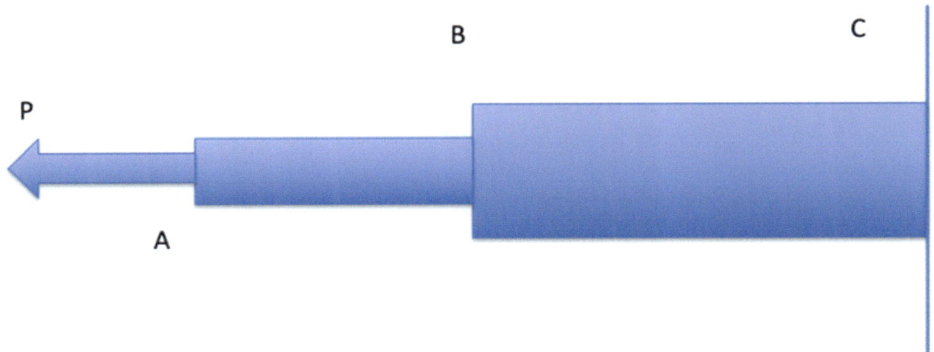

Fig. 2.8 Problem definition

```
% Definition of Stiffness Matrices
clear
L1 = 1;

 L1 = 1

L2 = 1.5;

 L2 = 1.5000

A1 = 200E-6;

 A1 = 2.0000e-04

A2 = 400E-6;

 A2 = 4.0000e-04

E1 = 200E9;

 E1 = 2.0000e+11

E2 = 100E9;

 E2 = 1.0000e+11

k(1) = A1*E1/L1

 k = 40000000
```

```
k(2) = A2*E2/L2

 k = 1×2
 10⁷ ×
     4.0000    2.6667


% Initializing
elem_con = [1 2; 2 3]; %element connectivity
elem = size(elem_con,1);%total nummber of elements
nodes = elem+1;% in a 1 D system #nodes = #elements+1
K = zeros(nodes);%Initialize global stiffness matrix



%Assembly
for i=1:elem
    ke = k(i)*[1 -1;-1 1]
    node1 = elem_con(i,1);
    node2 = elem_con(i,2);
    K(node1,node1) = K(node1,node1) + ke(1,1);
    K(node1,node2) = K(node1,node2) + ke(1,2);
    K(node2,node1) = K(node2,node1) + ke(2,1);
    K(node2,node2) = K(node2,node2) + ke(2,2);
end

 ke = 2×2
     40000000    -40000000
    -40000000     40000000
 ke = 2×2
 10⁷ ×
     2.6667    -2.6667
    -2.6667     2.6667


K

 K = 3×3
 10⁷ ×
     4.0000    -4.0000          0
    -4.0000     6.6667    -2.6667
          0    -2.6667     2.6667

P = -50E3;
%Forces
F = [P;0;0]
```

```
 F = 3×1
        -50000
             0
             0
```

```
d = zeros(nodes,1)
```

```
 d = 3×1
        0
        0
        0
```

```
% Solve the system of equations for the unknown displacements
d(1:2) = K(1:2,1:2)\F(1:2)
```

```
 d = 3×1
     -0.0031
     -0.0019
           0
```

```
% And use the displacement vector to calculate all the unknown forces
% Such as the reaction forces.
F = K*d
```

```
 F = 3×1
 10^4 ×
     -5.0000
      0.0000
      5.0000
```

```
% Use the displacement of each element or the end diplacement of the
two
% nodes to calculate the stresses in each element
del1 = d(1:2)
```

```
 del1 = 2×1
    -0.0031
    -0.0019
```

```
del2 = d(2:3)
```

```
 del2 = 2×1
     -0.0019
           0
```

```
% Calculating normal stresses in the two zones
Stress(1) = E1* (d(2)-d(1))/L1

  Stress = 1×2
  10⁸ ×
       2.5000     1.2500


Stress(2) = E2*(d(3)-d(2))/L2

  Stress = 1×2
  10⁸ ×
       2.5000     1.2500


plot(d,'LineWidth',3); % plotting the displacement at the nodes

%Plot properties
%legend('show')
title('Displacement ')
xlabel('Node')
ylabel('Horizontal displacement')


bar(Stress); %plotting the stresses in each element

%Plot properties
legend('show')
title('Stress in the elements ')
xlabel('Element')
ylabel('Stress')
```

Figure 2.9 shows the displacements at the nodes and Fig. 2.10 shows the stresses in each element.

2.2 Dynamic Problems

The simplest type of dynamic problem consists of a spring mass problem where we consider a set of springs and masses attached to each other such as the ones shown in Fig. 2.11. The most common analysis in this case is an analysis to determine the natural frequency and the mode shapes (also known as eigen values and eigenvectors of the system). The natural frequency and mode shapes can inform us about how the system will dynamically respond to external disturbances and which frequency contained in the disturbing force may evoke what type of response from the system.

To establish the process of analysis we consider a generic spring element like the one seen before with a mass at each node or end point (Fig. 2.12).

So, the earlier set of static equations can now be written as a set of dynamic equations:

$$\sum F_x = Ma \qquad (2.28)$$

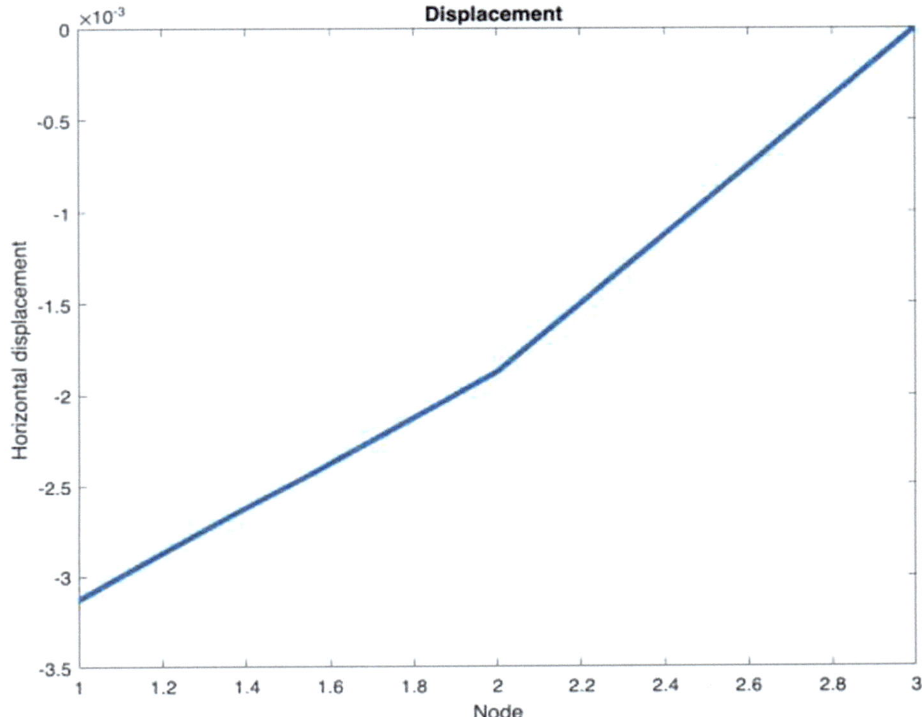

Fig. 2.9 Displacement profile

written at the two node this equation becomes:

$$f1_x + k(d2 - d1) = m1\,\ddot{d1} \tag{2.29}$$

$$f2_x + k(d1 - d2) = m2\,\ddot{d2} \tag{2.30}$$

or

$$f1_x = -k(d2 - d1) + m1\,\ddot{d1} \tag{2.31}$$

$$f2_x = -k(d1 - d2) + m2\,\ddot{d2} \tag{2.32}$$

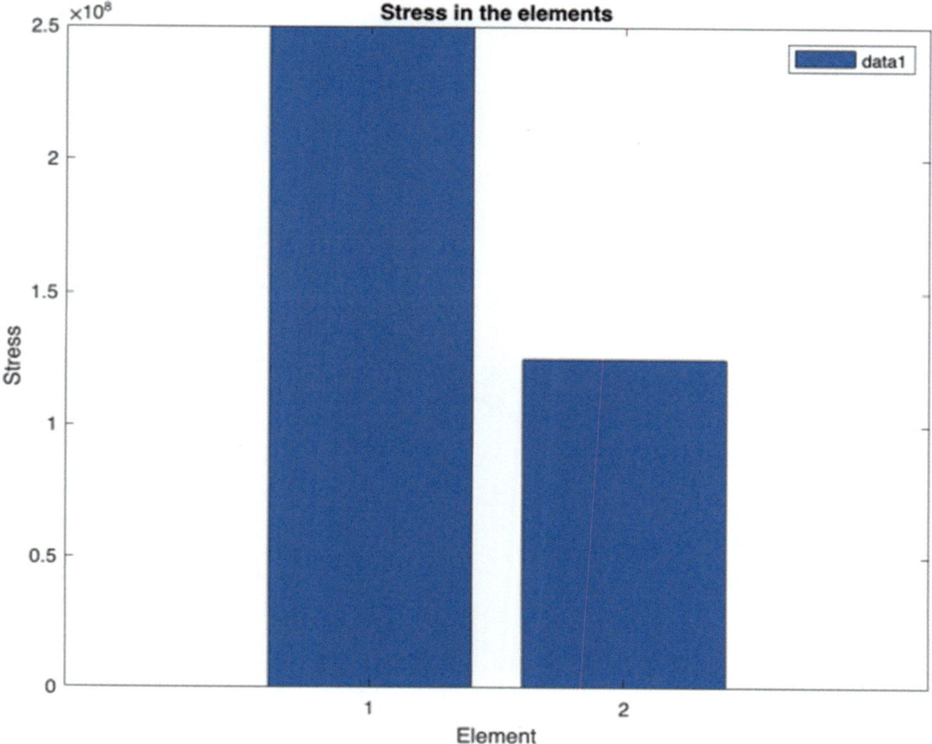

Fig. 2.10 Stresses in each element

This set of equations in matrix form is written as:

$$\begin{bmatrix} m1 & 0 \\ 0 & m2 \end{bmatrix} \begin{bmatrix} \ddot{d}_1 \\ \ddot{d}_2 \end{bmatrix} + \begin{bmatrix} k & -k \\ -k & k \end{bmatrix} \begin{bmatrix} d1 \\ d2 \end{bmatrix} = \begin{bmatrix} f1 \\ f2 \end{bmatrix} \qquad (2.33)$$

To solve these equations, when no forces are acting, initial conditions are required. This condition is known as the free vibration situation and the right-hand side is equal to zero. If we set the right-hand sides equal to zero:

$$\begin{bmatrix} m1 & 0 \\ 0 & m2 \end{bmatrix} \begin{bmatrix} \ddot{d}_1 \\ \ddot{d}_2 \end{bmatrix} + \begin{bmatrix} k & -k \\ -k & k \end{bmatrix} \begin{bmatrix} d1 \\ d2 \end{bmatrix} = \begin{bmatrix} 0 \\ 0 \end{bmatrix} \qquad (2.34)$$

The oscillatory nature of the solution requires the solution to be:

$$\begin{bmatrix} d1 \\ d2 \end{bmatrix} = \begin{bmatrix} D1 \\ D2 \end{bmatrix} \mathrm{Sin}\omega t \qquad (2.35)$$

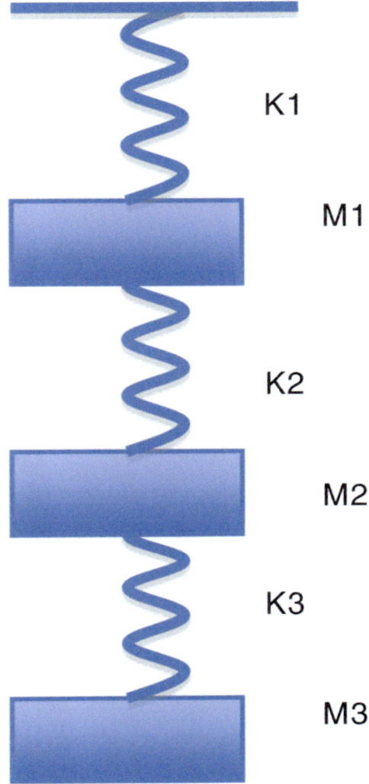

Fig. 2.11 A typical spring mass system

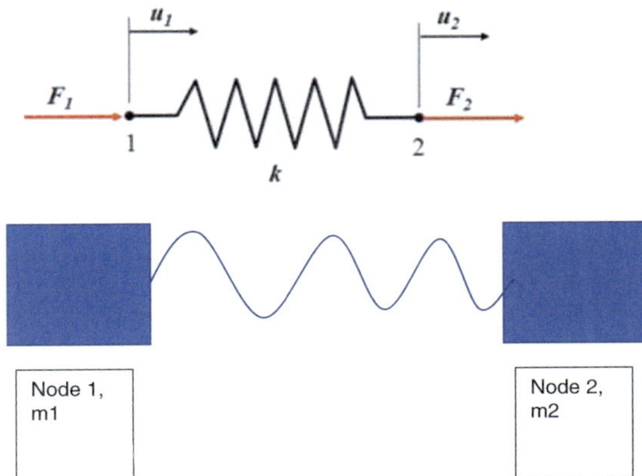

Fig. 2.12 A spring with two masses at the two end nodes

where Ds are the amplitudes and ω is the natural frequency. So,

$$\begin{bmatrix} \ddot{d1} \\ \ddot{d2} \end{bmatrix} = -\omega^2 \begin{bmatrix} D1 \\ D2 \end{bmatrix} Sin\omega t \qquad (2.36)$$

By substituting this solution in the system we get:

$$\left[\begin{bmatrix} k & -k \\ -k & k \end{bmatrix} - \omega^2 \begin{bmatrix} m1 & 0 \\ 0 & m2 \end{bmatrix} \right] \begin{bmatrix} D1 \\ D2 \end{bmatrix} = \begin{bmatrix} 0 \\ 0 \end{bmatrix} \qquad (2.37)$$

There is a trivial solution to this system which is that D1 and D2 are zeros. But for this system to have non-trivial solution the determinant of the matrix has to be equal to zero:

$$\left| \begin{bmatrix} k & -k \\ -k & k \end{bmatrix} - \omega^2 \begin{bmatrix} m1 & 0 \\ 0 & m2 \end{bmatrix} \right| = 0 \qquad (2.38)$$

Consider a problem where we have 2 springs and 2 masses. So, the above equation is written for one spring and a second can be written similarly. One end of the first spring is fixed so the two equations are written in assembled form as:

$$\left(\begin{bmatrix} k1 & -k1 & 0 \\ -k1 & k1+k2 & -k2 \\ 0 & -k2 & k2 \end{bmatrix} - \omega^2 \begin{bmatrix} 0 & 0 & 0 \\ 0 & m1 & 0 \\ 0 & 0 & m2 \end{bmatrix} \right) \begin{bmatrix} d1 \\ d2 \\ d3 \end{bmatrix} = 0 \qquad (2.39)$$

$$\left| \begin{bmatrix} k1 & -k1 & 0 \\ -k1 & k1+k2 & -k2 \\ 0 & -k2 & k2 \end{bmatrix} - \omega^2 \begin{bmatrix} 0 & 0 & 0 \\ 0 & m1 & 0 \\ 0 & 0 & m2 \end{bmatrix} \right| = 0 \qquad (2.40)$$

this results in Eq. 2.41 (since d1 = 0):

$$\left| \begin{bmatrix} k1+k2 & -k2 \\ -k2 & k2 \end{bmatrix} - \omega^2 \begin{bmatrix} m1 & 0 \\ 0 & m2 \end{bmatrix} \right| = 0 \qquad (2.41)$$

which is,

$$\left(\frac{(k1+k2)}{m1} - \omega^2 \right) \left(\frac{k2}{m2} - \omega^2 \right) - \frac{k2^2}{m1m2} = 0 \qquad (2.42)$$

Solving this for ω gives the two natural frequencies.

And substituting this in the matrix equation we get two sets of vectors for d2 and d3 which are known as eigen vectors.

Any displacement of the two masses can be expressed as a combination of the eigenvectors.

The eigenvectors are written as

$$[V] = [V1 \quad V2] = \begin{bmatrix} V1_1 & V2_1 \\ V1_2 & V2_2 \end{bmatrix} \tag{2.43}$$

So any displacement is written as (with a and b as unknown multipliers);

$$\begin{bmatrix} X1 \\ X2 \end{bmatrix} = aV1 + bV2 = a\begin{bmatrix} V1_1 \\ V1_2 \end{bmatrix} + b\begin{bmatrix} V2_1 \\ V2_2 \end{bmatrix} \tag{2.44}$$

we would like to find a and b for a given initial condition. So if X0 is a vector of specified initial conditions:

$$\begin{bmatrix} X0_1 \\ X0_2 \end{bmatrix} = aV1 + bV2 = a\begin{bmatrix} V1_1 \\ V1_2 \end{bmatrix} + b\begin{bmatrix} V2_1 \\ V2_2 \end{bmatrix} = [V1 \quad V2]\begin{bmatrix} a \\ b \end{bmatrix} \tag{2.45}$$

So,

$$\begin{bmatrix} a \\ b \end{bmatrix} = [V]^{-1}[X0] \tag{2.46}$$

After a and b are computed the system response using the eigenvectors can be computed as:

$$\begin{bmatrix} X1 \\ X2 \end{bmatrix} = aV1\cos\omega_1 t + bV2\cos\omega_2 t \tag{2.47}$$

So once the natural frequency and eigenvectors are found response with any initial input can be determined.

Example 2.5 We will consider an example with two springs and two masses with one (left) end of the first spring attached to the ground or wall so it is fixed.

We will modify the code from static problem discussion with spring elements to create the process of assembly and solution.

```matlab
clear
% if more springs are part of the system more k's
% will be added
% sliders can be used to alter the input variables
k(1) =18
```

k = 18

```matlab
k(2) =10
```

k = 1×2
     18   10

```matlab
% Definition of Stiffness Matrices
kel1 = k(1)*[1 -1;-1 1]
```

kel1 = 2×2
     18   -18
    -18   18

```matlab
kel2 = k(2)*[1 -1;-1 1]
```

kel2 = 2×2
     10   -10
    -10   10

```matlab
% if more masses are part of the system more m's
% will be added

m(1)=2
```

m = 2

```matlab
m(2)=1
```

m = 1×2
      2   1

```matlab
% Initializing
elem_con = [1 2; 2 3]; %elemment connectivity, more
% elements will be added with additional elements
elem = size(elem_con,1);%total nnummber of elements
nodes = elem+1; % in a 1 D system #nodes = #elements+1
K = zeros(nodes); %Initialize global stiffness matrix
%Assembly
for i=1:elem
    ke = k(i)*[1 -1;-1 1];
    node1 = elem_con(i,1);
```

```matlab
    node2 = elem_con(i,2);
    K(node1,node1) = K(node1,node1) + ke(1,1);
    K(node1,node2) = K(node1,node2) + ke(1,2);
    K(node2,node1) = K(node2,node1) + ke(2,1);
    K(node2,node2) = K(node2,node2) + ke(2,2);
end
%Divide both equations by the masses
for i=2:nodes
    K(i,1:nodes)=K(i,1:nodes)/m(i-1);
end
K
```

```
K = 3×3
    18    -18      0
    -9     14     -5
     0    -10     10
```

```matlab
%Boundary condition
% Node 1 is fixed
% So the system of equations is reduced
% this part will have to be altered if a different node is fixed
Kred = K(2:3,2:3)
```

```
Kred = 2×2
    14     -5
   -10     10
```

```matlab
% V is the Eigen vector matrix and D is the eigenvalue matrix
[V,D] = eig(Kred)
```

```
V = 2×2
    0.6829    0.4716
   -0.7305    0.8818
D = 2×2
   19.3485         0
         0    4.6515
```

```matlab
% D is the vector containing the eigenvalues.  Sqrt of the
eigenvalues
% are the   natural frequencies
omega = sqrt(eig(Kred))
```

```
omega = 2×1
    4.3987
    2.1567
```

We will now compute the dynamic behavior of the two masses for a given initial displacement condition (1, 0) (results shown in Fig. 2.13).

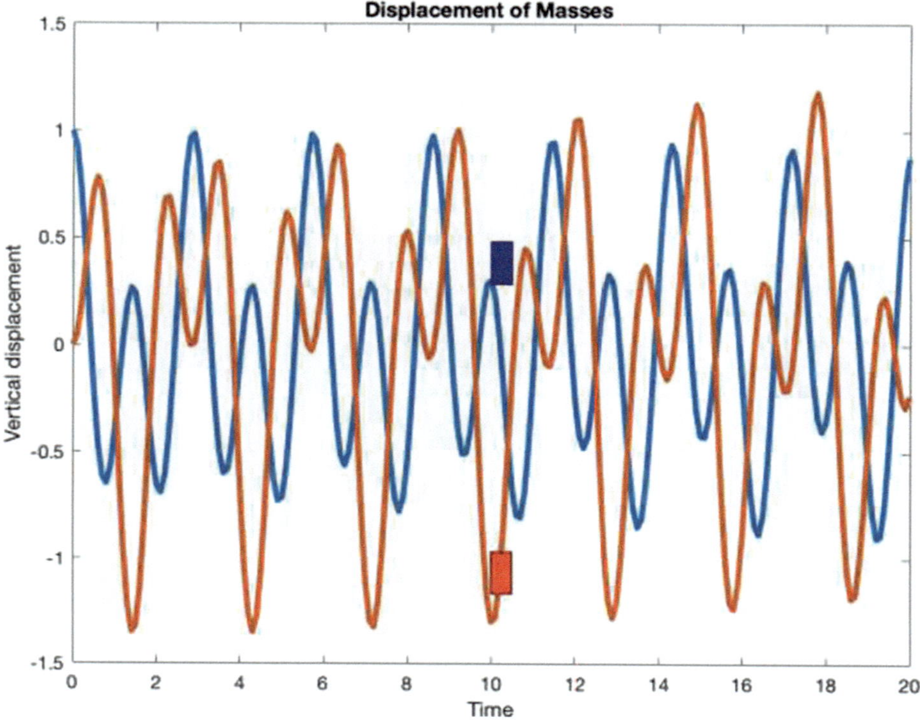

Fig. 2.13 Vertical displacement of Masses for assumed initial displacement

```matlab
% In this section we use the eigenvalues and eigenvectors to commpute the
% system response for any initial condition.
x0=[1;0] % initial conditions
```

```
x0 = 2×1
     1
     0
```

```matlab
% With an initial displacement vector of x0 we are going to compute the
% mmotion of the two masses using the equations shown before.
gam = inv(V)*x0
```

```
gam = 2×1
    0.9314
    0.7716
```

```matlab
g = diag(gam)
```

```
g = 2×2
    0.9314         0
         0    0.7716
```

```matlab
t = 0:0.1:20
```

```
t = 1×201
         0    0.1000    0.2000    0.3000    0.4000    0.5000
0.6000 ⋯
```

```matlab
x = V*g*cos(omega*t)
```

```
x = 2×201
    1.0000    0.9310    0.7360    0.4485    0.1174   -0.2020    -
0.4581 ⋯
         0    0.0490    0.1844    0.3738    0.5702    0.7217
0.7823
```

```matlab
plot(t,x,'LineWidth',3)
```

```matlab
%Animmation of the masses in motion
%This can be altered by changing the input vector x0 and recomputing this
%last section
```

```
data1 = [10 x0(1) 0.5 0.2];
data2 = [10 x0(2) 0.5 0.2];
hold on
r1 = rectangle ('position', data1,"FaceColor",[0 0 1]);
r2 = rectangle ('position', data2,"FaceColor",[1 0 0]);
hold off
for j = 1:1:100
    data1(2)= x(1,j);
    data2(2)= x(2,j);
    r1.Position = data1;
    r2.Position = data2;
    drawnow
end
title('Displacement of Masses ')
xlabel('Time')
ylabel('Vertical displacement')
```

Example 2.6 The following figure shows a schematic that represents a 2-story building. The vertical members supporting the masses can be treated as springs and each floor can be considered a mass. These structures are capable of going through horizontal vibration.

We will find the **natural frequency and the mode shapes** for this system. **K1 = 120 kN/m, K2 = 120 kN/m. M1 = 500 kg, M2 = 300 kg** (Fig. 2.14).

We will now compute the dynamic behavior of the two masses for a given initial displacement condition.

Fig. 2.14 A structure replicating two floors of a building

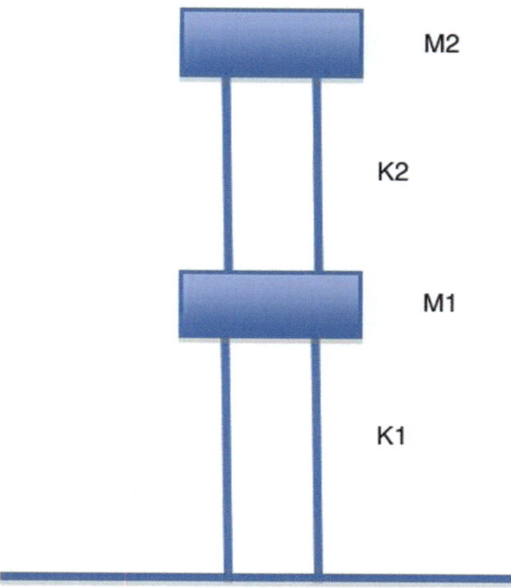

```
clear
k(1) = 120E3

k = 120000

k(2) = 120E3

k = 1×2


K

K = 3×3
      120000       -120000             0
        -240           480          -240
           0          -400           400

%Boundary condition
% Node 1 is fixed
% So the system of equations is reduced
% this part will have to be altered if a different node is fixed
Kred = K(2:3,2:3)

Kred = 2×2
   480   -240
  -400    400

% V is the Eigen vector matrix and D is the eigenvalue matrix
[V,D] = eig(Kred)

V = 2×2
    0.6611      0.5629
   -0.7503      0.8265
D = 2×2
  752.4100           0
         0    127.5900

% D is the vector containing the eigenvalues.  Sqrt of the
eigenvalues
% are the  natural frequencies
omega = sqrt(eig(Kred))

omega = 2×1
   27.4301
   11.2956
```

```
        120000          120000
```

```
% Definition of Stiffness Matrices
kel1 = k(1)*[1 -1;-1 1]
```

```
kel1 = 2×2
        120000         -120000
       -120000          120000
```

```
kel2 = k(2)*[1 -1;-1 1]
```

```
kel2 = 2×2
        120000         -120000
       -120000          120000
```

```
% if more masses are part of the system more m's
% will be added
```

```
m(1)= 500
```

```
m = 500
```

```
m(2)= 300
```

```
m = 1×2
    500     300
```

```
% Initializing
elem_con = [1 2; 2 3]; %elemment connectivity, more
% elements will be added with additional elements
elem = size(elem_con,1);%total nnummber of elements
nodes = elem+1; % in a 1 D system #nodes = #elements+1
K = zeros(nodes); %Initialize global stiffness matrix
%Assembly
for i=1:elem
    ke = k(i)*[1 -1;-1 1];
    node1 = elem_con(i,1);
    node2 = elem_con(i,2);
    K(node1,node1) = K(node1,node1) + ke(1,1);
    K(node1,node2) = K(node1,node2) + ke(1,2);
    K(node2,node1) = K(node2,node1) + ke(2,1);
    K(node2,node2) = K(node2,node2) + ke(2,2);
end
%Divide both equations by the masses
for i=2:nodes
    K(i,1:nodes)=K(i,1:nodes)/m(i-1);
end
```

We will now compute the dynamic behavior of the two masses for a given initial displacement condition (0, 0.01). Results shown in Fig. 2.15)

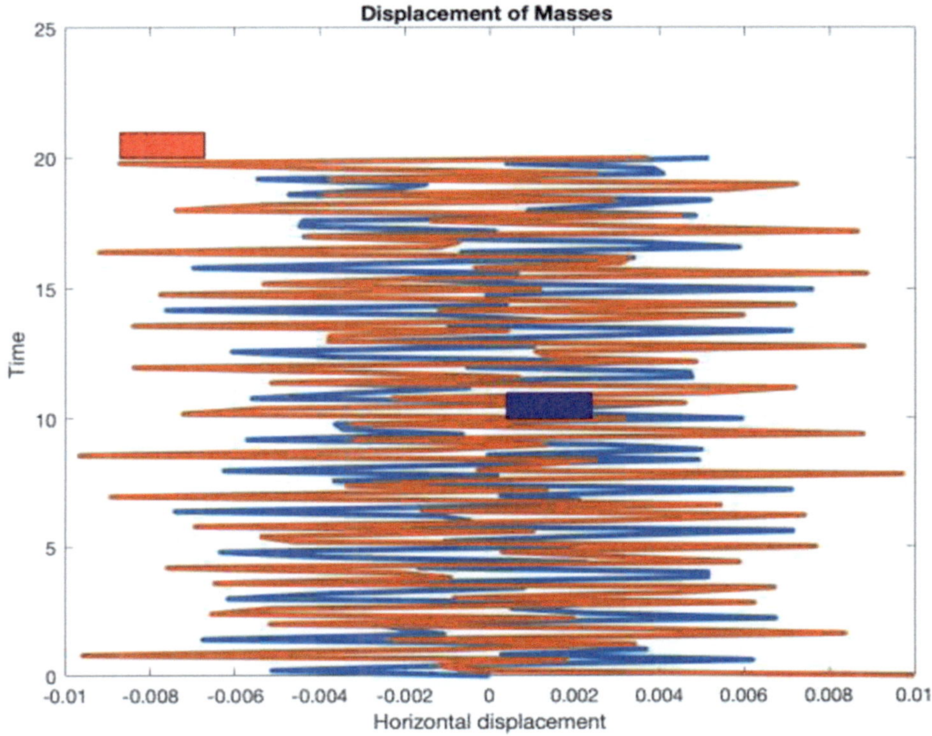

Fig. 2.15 Horizontal displacement of masses for assumed initial displacement

```
% In this section we use the eigenvalues and eigenvectors to commpute the
% system response for any initial condition.
x0=[0;0.01] % initial conditions

x0 = 2×1
         0
    0.0100
```

```
% With an initial displacement vector of x0 we are going to compute
the
% mmotion of the two masses using the equations shown before.
 gam = inv(V)*x0
```

gam = 2×1
```
   -0.0058
    0.0068
```

```
 g = diag(gam)
```

g = 2×2
```
   -0.0058        0
        0    0.0068
```

```
 t = 0:0.2:20
```

t = 1×101
```
        0    0.2000    0.4000    0.6000    0.8000    1.0000
1.2000 ...
```

```
 x = V*g*cos(omega*t)
```

x = 2×101
```
        0   -0.0051   -0.0007    0.0062    0.0003    0.0037
0.0018 ...
    0.0100   -0.0005   -0.0012    0.0018   -0.0096   -0.0012
0.0034
```

```
 plot(x,t,'LineWidth',3)

 %Animmation of the masses in motion
 %This can be altered by changing the input vector x0 and recomputing
this
 %last section

 data1 = [x0(1) 10 0.002 1.0];
 data2 = [x0(2) 20 0.002 1.0];
 hold on
 r1 = rectangle ('position', data1,"FaceColor",[0 0 1]);
 r2 = rectangle ('position', data2,"FaceColor",[1 0 0]);
 hold off
 for j = 1:1:100
     data1(1)= x(1,j);
     data2(1)= x(2,j);
     r1.Position = data1;

     r2.Position = data2;
     drawnow
 end
 title('Displacement of Masses ')
 xlabel('Horizontal displacement')
 ylabel('Time')
```

2.3   2D Truss Problems

Truss elements are similar to the linear spring elements but they can be oriented in all different directions in space. As a result, in a truss structure, all elements will not be oriented in the same fashion. To handle this we need to be aware of two main issues: (a) at every node the forces and displacements will need to be considered in two perpendicular directions, e.g. horizontal and vertical or along the length of the member and perpendicular to it, and (b) system(s) of co-ordinates need to be properly defined (Fig. 2.16).

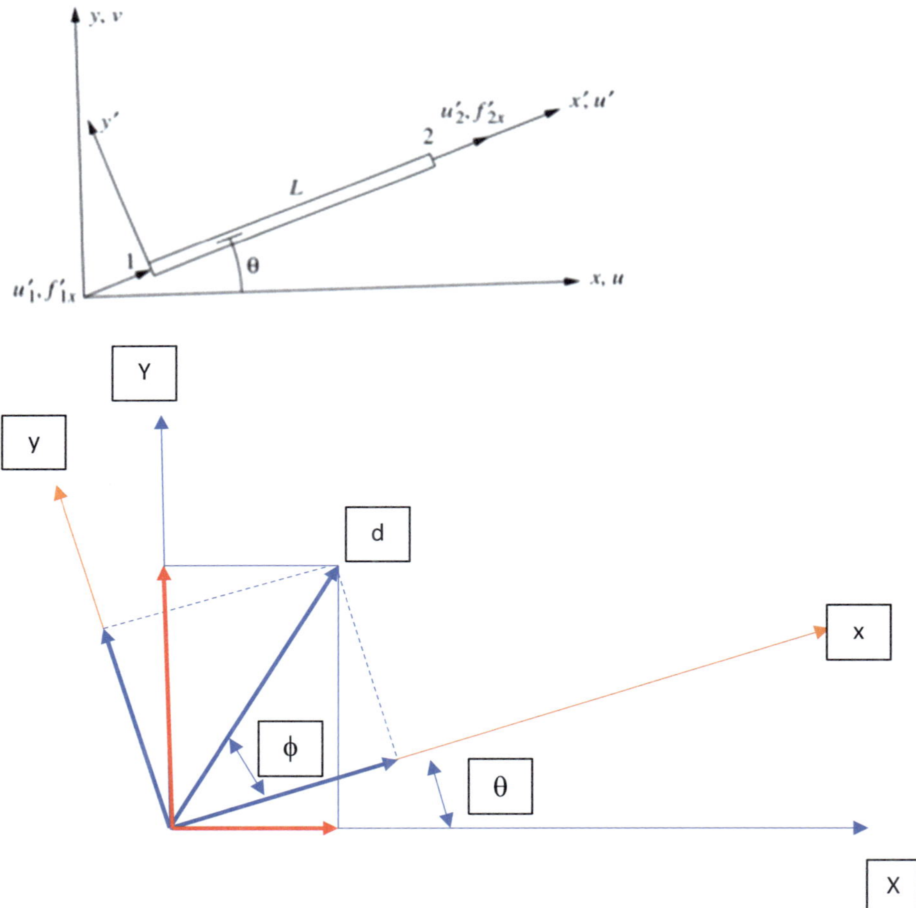

Fig. 2.16 Generic Truss Element

To establish this, we will first define a global co-ordinate system and a local co-ordinate system. The global system is the traditional ground-fixed Cartesian system with a horizontal X-axis and a vertical Y-axis. The local- or body fitted co-ordinate system is fixed to a particular truss member. Figure 2.16 shows the local and global co-ordinates.

X and Y represent the global co-ordinate system and x and y represent the local co-ordinate system. Angle θ measured positive counter-clockwise going from global X to local x is the angle between the two co-ordinate systems. Let us consider any vector **d** such that it makes an angle ϕ with the local co-ordinate system and as shown in the figure it makes $(\theta + \phi)$ with the global co-ordinate system.

So, the two components of the vector in the local co-ordinate system are:

$$
\begin{aligned}
d_x &= d\mathrm{Cos}\phi \\
d_y &= d\mathrm{Sin}\phi
\end{aligned}
\tag{2.48}
$$

And the two components in the global co-ordinate system are:

$$
\begin{aligned}
d_X &= d\mathrm{Cos}(\phi + \theta) = d\mathrm{Cos}\phi\mathrm{Cos}\theta - d\mathrm{Sin}\phi\mathrm{Sin}\theta = d_x\mathrm{Cos}\theta - d_y\mathrm{Sin}\theta \\
d_Y &= d\mathrm{Sin}(\phi + \theta) = d\mathrm{Sin}\phi\mathrm{Cos}\theta + d\mathrm{Sin}\theta\mathrm{Cos}\phi = d_y\mathrm{Cos}\theta + d_x\mathrm{Sin}\theta
\end{aligned}
\tag{2.49}
$$

Hence, we can write:

$$
\begin{bmatrix} d_X \\ d_Y \end{bmatrix} = \begin{bmatrix} c & -s \\ s & c \end{bmatrix} \begin{bmatrix} d_x \\ d_y \end{bmatrix}
\tag{2.50}
$$

where c means $\mathrm{Cos}\theta$ and s means $\mathrm{Sin}\theta$. The equation is written for any vector d. For our consideration here the vector could be displacement or force. So for any force vector for each node the equations can be written as:

$$
\begin{bmatrix} F1_X \\ F1_Y \\ F2_X \\ F2_Y \end{bmatrix} = \begin{bmatrix} c & -s & 0 & 0 \\ s & c & 0 & 0 \\ 0 & 0 & c & -s \\ 0 & 0 & s & c \end{bmatrix} \begin{bmatrix} f1_x \\ f1_y \\ f2_x \\ f2_y \end{bmatrix}
\tag{2.51}
$$

and for a displacement vector at each node, it can be written as:

$$
\begin{bmatrix} d1_X \\ d1_Y \\ d2_X \\ d2_Y \end{bmatrix} = \begin{bmatrix} c & -s & 0 & 0 \\ s & c & 0 & 0 \\ 0 & 0 & c & -s \\ 0 & 0 & s & c \end{bmatrix} \begin{bmatrix} d1_x \\ d1_y \\ d2_x \\ d2_y \end{bmatrix}
\tag{2.52}
$$

we can write this transformation as:

$$
[D] = [T][d]
\tag{2.53}
$$

or

$$[d] = [T]^{-1}[D] \tag{2.54}$$

and

$$[F] = [T][f] \tag{2.55}$$

$$[f] = [T]^{-1}[F] \tag{2.56}$$

where T is the transformation matrix.

For the strut or spring element using the local coordinate system:

$$\begin{bmatrix} f1_x \\ f2_x \end{bmatrix} = \frac{AE}{L} \begin{bmatrix} 1 & -1 \\ -1 & 1 \end{bmatrix} \begin{bmatrix} d1_x \\ d2_x \end{bmatrix} \tag{2.57}$$

Including both components of force and displacement this can be written as:

$$\begin{bmatrix} f1_x \\ f1_y \\ f2_x \\ f2_y \end{bmatrix} = \frac{AE}{L} \begin{bmatrix} 1 & 0 & -1 & 0 \\ 0 & 0 & 0 & 0 \\ 1 & 0 & -1 & 0 \\ 0 & 0 & 0 & 0 \end{bmatrix} \begin{bmatrix} d1_x \\ d1_y \\ d2_x \\ d2_y \end{bmatrix} \tag{2.58}$$

So, in local co-ordinates this relationship can be written as:

$$[f] = [K][d] \tag{2.59}$$

Substituting the global quantities we get

$$[f] = [T]^{-1}[F] = [K][d] = [K][T]^{-1}[D] \tag{2.60}$$

$$[T]^{-1}[F] = [K][T]^{-1}[D] \tag{2.61}$$

$$[F] = [T][K][T]^{-1}[D] \tag{2.62}$$

For this system the standard stiffness matrix for the element is written as

$$[T][K][T]^{-1} = \frac{AE}{L} \begin{bmatrix} c^2 & cs & -c^2 & -cs \\ cs & s^2 & -cs & -s^2 \\ -c^2 & -cs & c^2 & cs \\ -cs & -s^2 & cs & s^2 \end{bmatrix} \tag{2.63}$$

When a truss-problem is to be solved matrices such as this one need to be assembled for each element. Here the letters c and s stand for cosine and sine of θ respectively.

This global stiffness matrix for the trusses relates the forces in global X and Y directions at the two end nodes of the element with the displacements as shown in Eq. 2.64

$$
\begin{bmatrix} F1_X \\ F1_Y \\ F2_X \\ F2_Y \end{bmatrix} = \frac{AE}{L} \begin{bmatrix} c^2 & cs & -c^2 & -cs \\ cs & s^2 & -cs & -s^2 \\ -c^2 & -cs & c^2 & cs \\ -cs & -s^2 & cs & s^2 \end{bmatrix} \begin{bmatrix} d1_X \\ d1_Y \\ d2_X \\ d2_Y \end{bmatrix}
\tag{2.64}
$$

Finding Stresses in the Truss elements

Finding stresses and strains in each element. The stresses and strains in each element are more easily calculated using the deflections in the local co-ordinate system. In fact, in a local co-ordinate system that is aligned with the direction of the truss element itself, the strain can be written as:

$$
\epsilon = \frac{(d2_x - d1_x)}{L}
\tag{2.65}
$$

In the local coordinate systems:

$$
\begin{bmatrix} f1_x \\ f2_x \end{bmatrix} = \frac{AE}{L} \begin{bmatrix} 1 & -1 \\ -1 & 1 \end{bmatrix} \begin{bmatrix} d1_x \\ d2_x \end{bmatrix}
\tag{2.66}
$$

So,

$$
f2_x = \frac{AE}{L} \{d2_x - d1_x\}
\tag{2.67}
$$

$$
\sigma = \frac{f2_x}{A} = \frac{E}{L} \{d2_x - d1_x\} = \frac{E}{L} [-1 \quad 1] \begin{bmatrix} d1_x \\ d2_x \end{bmatrix} = \frac{E}{L} [-1 \quad 1][d]
\tag{2.68}
$$

In global coordinates this can be written as:

$$
\sigma = \frac{E}{L} [-1 \quad 1][T]^{-1}[D]
\tag{2.69}
$$

```
syms c
syms s
T = [c -s 0 0;s c 0 0;0 0 c -s;0 0 s c]

T =

 ⎛ c  -s   0   0 ⎞
 ⎜ s   c   0   0 ⎟
 ⎜ 0   0   c  -s ⎟
 ⎝ 0   0   s   c ⎠

 inv(T)

ans =
```

$$\begin{pmatrix} \sigma_1 & \frac{s}{c^2+s^2} & 0 & 0 \\ -\frac{s}{c^2+s^2} & \sigma_1 & 0 & 0 \\ 0 & 0 & \sigma_1 & \frac{s}{c^2+s^2} \\ 0 & 0 & -\frac{s}{c^2+s^2} & \sigma_1 \end{pmatrix}$$

where

$$\sigma_1 = \frac{s}{c^2+s^2}$$

$$[T]^{-1} = \begin{bmatrix} c & s & 0 & 0 \\ -s & c & 0 & 0 \\ 0 & 0 & c & s \\ 0 & 0 & -s & c \end{bmatrix} \tag{2.70}$$

$$\sigma = \frac{E}{L}[-1 \quad 1][T]^{-1}[D] = \frac{E}{L}[-1 \quad 1]\begin{bmatrix} c & s & 0 & 0 \\ 0 & 0 & c & s \end{bmatrix}\begin{bmatrix} d1_X \\ d1_Y \\ d2_X \\ d2_Y \end{bmatrix}$$

$$= \frac{E}{L}[-c \quad -s \quad c \quad s]\begin{bmatrix} d1_X \\ d1_Y \\ d2_X \\ d2_Y \end{bmatrix} \tag{2.71}$$

Example 2.7 Consider the truss shown in the Fig. 2.17. Let E = 210 GPa and all the cross-sectional areas are 6E-4 m^4. A force of 1000 KN acts as shown in the figure. The structure is made of two truss elements 1 and 2 as shown in the figure. Nodes 2 and 3 are pinned down such that no movement in the x or y directions are allowed at those locations (Fig. 2.17).

Fig. 2.17 Schematic for Example 2.7

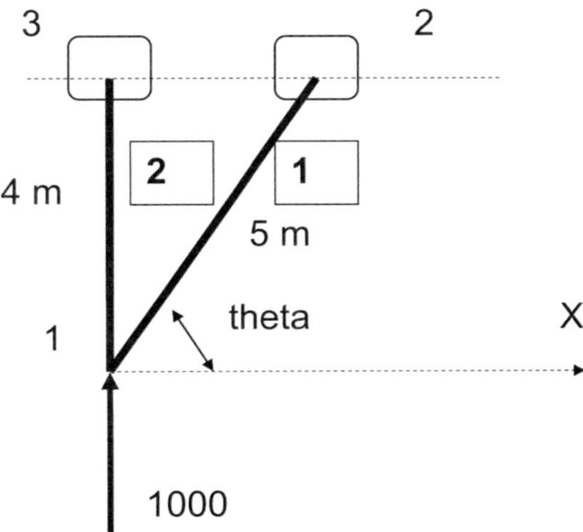

For Element #1

Element #1 c = 3/5 = 0.6, s = 4/5 = 0.8

$$\begin{bmatrix} F1_X \\ F1_Y \\ F2_X \\ F2_Y \end{bmatrix} = \frac{(6E-4)(210E9)}{5} \begin{bmatrix} 0.36 & 0.48 & -0.36 & -0.48 \\ 0.48 & 0.64 & -0.48 & -0.64 \\ -0.36 & -0.48 & 0.36 & 0.48 \\ -0.48 & -0.64 & 0.48 & 0.64 \end{bmatrix} \begin{bmatrix} d1_X \\ d1_Y \\ d2_X \\ d2_Y \end{bmatrix} \qquad (2.72)$$

For Element 2

Element #2 c = 0, s = 1

$$\begin{bmatrix} F1_X \\ F1_Y \\ F3_X \\ F3_Y \end{bmatrix} = \frac{(6E-4)(210E9)}{4} \begin{bmatrix} 0 & 0 & 0 & 0 \\ 0 & 1 & 0 & -1 \\ 0 & 0 & 0 & 0 \\ 0 & -1 & 0 & 1 \end{bmatrix} \begin{bmatrix} d1_X \\ d1_Y \\ d3_X \\ d3_Y \end{bmatrix} \qquad (2.73)$$

After these two equations assembled the combined equation for the whole structure looks like

$$\begin{bmatrix} F1_X \\ F1_Y \\ F2_X \\ F2_Y \\ F3_X \\ F3_Y \end{bmatrix} = \frac{(6E-4)(210E9)}{5} \begin{bmatrix} 0.36 & 0.48 & -0.36 & -0.48 & 0 & 0 \\ 0.48 & 0.64 & -0.48 & -0.64 & 0 & -1.25 \\ -0.36 & -0.48 & 0.36 & 0.48 & 0 & 0 \\ -0.48 & -0.64 & 0.48 & 0.64 & 0 & 0 \\ 0 & 0 & 0 & 0 & 0 & 0 \\ 0 & -1.25 & 0 & 0 & 0 & 1.25 \end{bmatrix} \begin{bmatrix} d1_X \\ d1_Y \\ d2_X \\ d2_Y \\ d3_X \\ d3_Y \end{bmatrix}$$

$$(2.74)$$

The boundary and force conditions are:

F1Y = 1000 KN

d2x = d2y = d3x = d3y = 0.0

By applying these conditions, we get a set of reduced equations:

$$\begin{bmatrix} F1_X \\ F1_Y \end{bmatrix} = 25{,}200 \begin{bmatrix} 0.36 & 0.48 \\ 0.4 & 1.89 \end{bmatrix} \begin{bmatrix} d1_X \\ d1_Y \end{bmatrix} \qquad (2.75)$$

```
K =25200*[0.36 0.48;0.48 1.89]

K = 2×2
        9072            12096
        12096           47628


F = [0;1000]

F = 2×1
             0
          1000


d = K\F

d = 2×1
     -0.0423
      0.0317
```

Solving these by MATLAB or otherwise the deflections are:

$$\begin{bmatrix} d1_X \\ d1_Y \end{bmatrix} = \begin{bmatrix} -4.23E - 2 \\ 3.174E - 2 \end{bmatrix} m \tag{2.76}$$

```
K = 25200*[-0.36 -0.48;-0.48 -0.64;0 0;0 -1.25]

K = 4×2
        -9072          -12096
       -12096          -16128
            0               0
            0          -31500


d

d = 2×1
     -0.0423
      0.0317


F = K*d

F = 4×1
10³ ×
      -0.0000
-0.0000
           0
-1.0000
```

Once the deflections are calculated they can be substituted back into the main system of equations to obtain the forces at the nodes 2 and 3 as:

$$
\begin{bmatrix} F2_X \\ F2_Y \\ F3_X \\ F3_Y \end{bmatrix} = \begin{bmatrix} 0 \\ 0 \\ 0 \\ -1000 \end{bmatrix} \text{KN}
\tag{2.77}
$$

Now we will do the same problem using actual code so that the creation of the matrices, the assembly, application of boundary conditions all can be automated (Fig. 2.18 shows the truss).

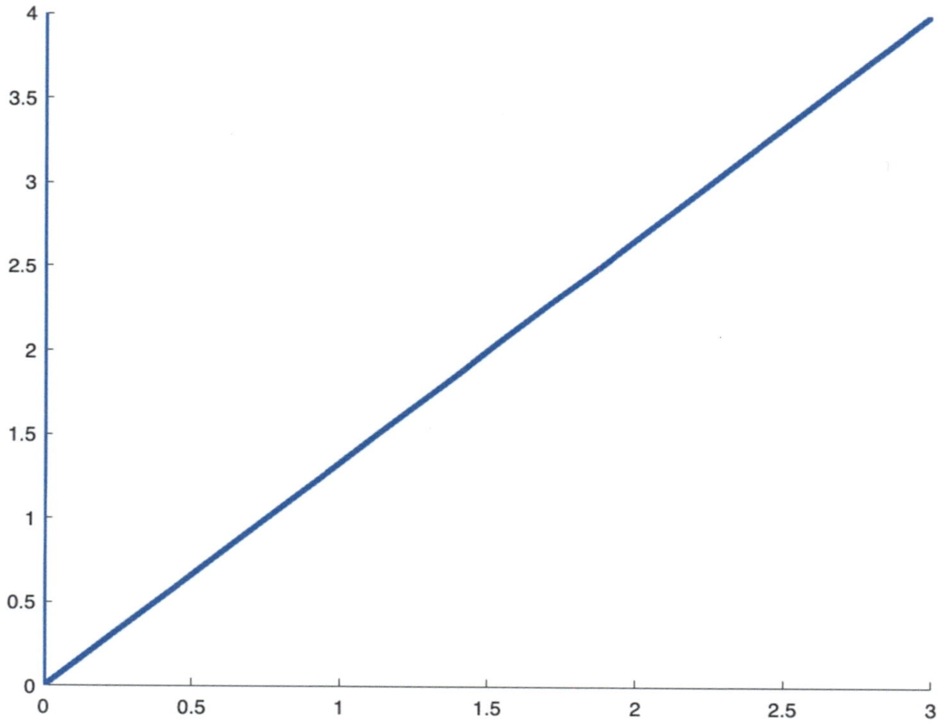

Fig. 2.18 2-element truss

```
% This program is written to solve 2-D Truss problems
clear
clc

% Following are the  different inputs sought: node,conn
(connectivity) and isol
% node- This variable holds the nodal co-ordinates of all corners of
the
% truss. It is provided in the form of a vector
% for example [x1 y1;x2 y2;x3 y3;so on...]

node = [0 0;3 4;0 4];


% conn- This variable stores the connectivity of the truss elemments
% that is which node is connected to which other node.  This is also
% provided in the form of a vector
% for example [1 2;1 3;2 3;2 4;1 4;3 4;3 6;4 5;4 6;3 5;5 6]
% Provide the starting node and end node of each element in the
structure.

conn = [1 2;1 3];

% isol- Each nodal point has two degrees of freedom, displacement in
the X direction
% and displacement in the  y direction.  Based on the support/
boundary conditions
% some of these degrees of freeedom are restricted.  This variable,
isol, lists
% the conditions that are free.  So if node number 1 is free to move
in
```

```
% either direction we will include 1 2 on this list.  If node 2 is
free to
% move in Y direction but not in X then wee will include 4 in this
list
% but not 3, etc.

isol = [1 2];
%A, cross-sectional area of the truss elements
A = 6E-4;
%E, Young's Mmodulus of the truss elements
E = 210E9;

%  The problem size, i.e size of the matrices (stiffness, load and
%  displacement) are determined from the provided data on element and
node
%  numbers
nn=size(node,1);
ndof=2*nn;        %Every node has 2 dofs.
ne=size(conn,1);         %this defines the size of elemental matrix
% Assign nodal connections and draw the structure
for e=1:ne
    n1=conn(e,1);        %  first node for the element
    n2=conn(e,2);        % second node for the element

    x1=node(n1,1);       % X1 co-ordinate of the first node
    y1=node(n1,2);       % Y1 co-ordinate of the first node
    x2=node(n2,1);       % X2 co-ordinate of the second node
    y2=node(n2,2);       % Y2 co-ordinate of the second node

X = [x1,x2];
Y = [y1,y2];
line(X,Y,'linewidth',3);
end
```

```
%The force vector is defined using the data available from the
problem.
%First the vector is initialized to all zero.  Then the non-zero
values are
%added
F=zeros(2*nn,1);
F(1)=0;
F(2)=1000E3;
F

F = 6×1
                0
          1000000
                0
                0
                0
                0
```

```matlab
% Initalization of the Stiffness mmatrix and the displacemment vector
K=zeros(ndof,ndof);      %Defines size of the global stiffness matrix
d=zeros(ndof,1);         %Defines size of the displacemetn matrix
% This loop is executed for all the elements. The loop calculates
%element lengths and the cosine and sine of its angular orientation
% then it  creates  the stiffness mmatrix for each elemment in the
variable
% ke.  Finally the global stiffness matrix is assembled.
 for e=1:ne
    n1=conn(e,1);         %  first node for the element
    n2=conn(e,2);         % second node for the element

    x1=node(n1,1);        % X1 co-ordinate of the first node
    y1=node(n1,2);        % Y1 co-ordinate of the first node
    x2=node(n2,1);        % X2 co-ordinate of the second node
    y2=node(n2,2);        % Y2 co-ordinate of the second node

    L=sqrt(((x2-x1))^2+((y2-y1)^2)); % Length of the truss element

    c=(x2-x1)/L;       %cosine of element orientation
    s=(y2-y1)/L;       %sine of element orientation

    % ke is the element matrix and all its members are populated
    ke=((A*E)/L)*[c*c,s*c,-c*c,-s*c;s*c,s*s,-s*c,-s*s;-c*c,-
s*c,c*c,s*c;-s*c,-s*s,s*c s*s];

    % Here, we globalize the global matrix according to the
respective node
    % and respective displacement
    sctr=[2*n1-1 2*n1 2*n2-1 2*n2];
    K(sctr,sctr)=K(sctr,sctr)+ke;
 end
 K

K = 6×6
10^7 ×
    0.9072      1.2096     -0.9072     -1.2096          0          0
    1.2096      4.7628     -1.2096     -1.6128          0    -3.1500
   -0.9072     -1.2096      0.9072      1.2096          0          0
   -1.2096     -1.6128      1.2096      1.6128          0          0
         0           0           0           0          0          0
         0     -3.1500           0           0          0     3.1500

% We have got our global matrix. Now we solve the matrices to get the
```

```matlab
% displacements at each node
d(isol)=K(isol,isol)\F(isol);
fprintf('\n----------Nodal Displacements----------');
```

----------Nodal Displacements----------

```matlab
fprintf('\nNo.  X-Direction       Y-Direction');
```

No. X-Direction   Y-Direction

```matlab
for i=1:nn
    fprintf('\n%5d  %8.3e    %8.3e',i,d(2*i-1),d(2*i));
end
```

```
1 -4.233e-02  3.175e-02
2  0.000e+00  0.000e+00
3  0.000e+00  0.000e+00
```

```matlab
%Once all the displacements are found the reactions at each node are
%calculated

f=K*d;
fprintf('\n----------Reactions----------');
```

----------Reactions----------

```matlab
fprintf('\nNo.  X-Direction       Y-Direction');
```

No. X-Direction   Y-Direction

```matlab
for i=1:nn
    fprintf('\n%5d  %8.3e    %8.3e',i,f(2*i-1),f(2*i));
end
```

```
1  0.000e+00  1.000e+06
2  0.000e+00  0.000e+00
3  0.000e+00 -1.000e+06
```

```matlab
% superpose the displaced structure on the original structure
for e=1:ne

    n1=conn(e,1);
    n2=conn(e,2);

    x1=node(n1,1);
    y1=node(n1,2);
    x2=node(n2,1);
```

```
    y2=node(n2,2);
    x1new = node(n1,1)+d(2*n1-1);
    y1new = node(n1,2)+d(2*n1);
    x2new = node(n2,1)+d(2*n2-1);
    y2new = node(n2,2)+d(2*n2);

    X = [x1new,x2new];
Y = [y1new,y2new];
line(X,Y,'linewidth',3,'color',[1 0 1]);
end
```

Figure 2.19 shows the displaced and the original structure.

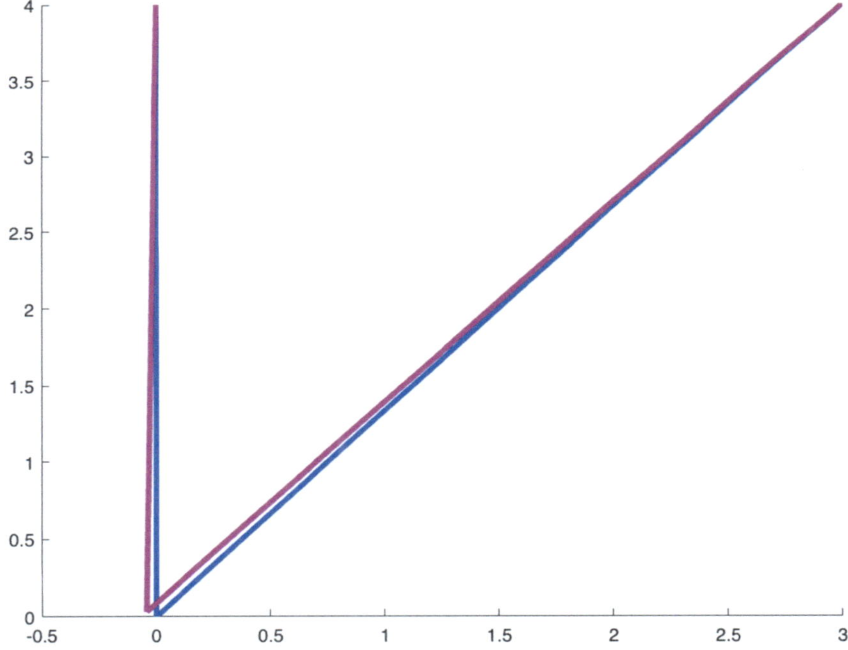

Fig. 2.19 Original and displaced truss

```matlab
% Now we print the strains and stresses
fprintf('\n----------Elemental strain & stress----------');
```

----------Elemental strain & stress----------

```matlab
fprintf('\nNo.  Strain  Stress');
```

No. Strain Stress

```matlab
for e=1:ne

    n1=conn(e,1);
    n2=conn(e,2);

    x1=node(n1,1);
    y1=node(n1,2);
    x2=node(n2,1);
    y2=node(n2,2);

    L=sqrt(((x2-x1))^2+((y2-y1)^2));
    c=(x2-x1)/L;
    s=(y2-y1)/L;

    % By following the standard calculations the strains and stresses
are
    % calculated
    B=(1/L)*[-c -s c s];

    % Again it is calculated by selecting the respective values from
the
    % globalized stiffness matrix
    sctr=[2*n1-1 2*n1 2*n2-1 2*n2];
    Strain=B*d(sctr);
    Stress=E*Strain;

    fprintf('\n%5d  %8.3e   %8.3e',e,Strain,Stress);
end
```

```
  1 -8.674e-19  -1.821e-07
  2 -7.937e-03  -1.667e+09
```

Example 2.8 Consider the truss problem shown in Fig. 2.20:

Let us create the input data to solve this problem.

node = [0 0;20 0;5 5;10 5;15 5;5 0;10 0;15 0].

conn = [1 3;1 6;6 3;6 4;6 7;3 4;7 4;7 8;4 8;4 5;8 5;8 2;5 2].

isol = [5 6 7 8 9 10 11 12 13 14 15 16].

A = 1500E-6.

E = 200E9.

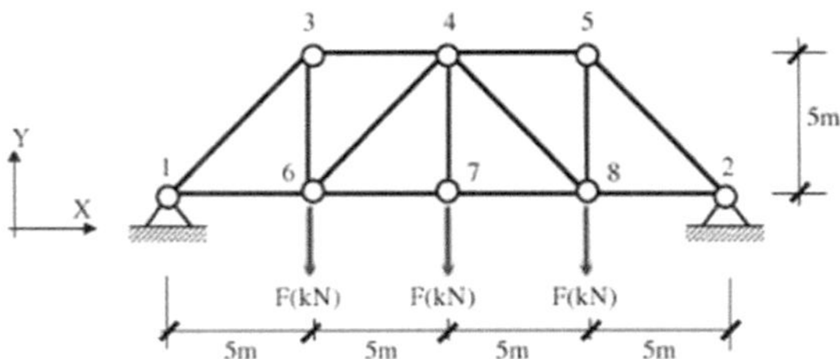

Fig. 2.20 Sketch for Example 2.8

F (12) = −1000E3.
F(14) = −1000E3.
F(16) = −1000E3.

```
% This program is written to solve 2-D  Truss problems
clear
clc

%Following are the  different inputs sought: node,conn (connectivity)
and isol
% node- This variable holds the nodal co-ordinaated of all corners of
the
% truss. It is provided iin the form of a vector
% for example [x1 y1;x2 y2;x3 y3;so on...]

node = [0 0;20 0;5 5;10 5;15 5;5 0;10 0;15 0];
```

```
% conn- This variable stores the connectivity of the truss elemments
% that is which node is connected to which other node.  This is also
% provided in the form of a vector
% for example [1 2;1 3;2 3;2 4;1 4;3 4;3 6;4 5;4 6;3 5;5 6]
% Provide the starting node and end node of each element in the
structure.

 conn =[1 3;1 6;6 3;6 4;6 7;3 4;7 4;7 8;4 8;4 5;8 5;8 2;5 2];

% isol- Each nodal point has two degrees of freedom, displacement in
the X direction
% and displacement in the  y direction.  Based on the support/
boundary conditions
% some of these degrees of freeedom are restricted.   This variable,
isol, lists
% the conditions that are free.  So if node number 1 is free to move
in
% either direction we will include 1 2 on this list.  If node 2 is
free to
% move in Y direction but not in X then wee will include 4 in this
list
% but not 3, etc.

 isol = [5 6 7 8 9 10 11 12 13 14 15 16];
%A, cross-sectional area of the truss elements
 A = 1500E-6;
%E, Young's Mmodulus of the truss elements
 E = 200E9;

%  The problem size, i.e size of the mmatrices (stiffness, load and
%  displacement) are determined from the provided data on element and
node
%  numbers
nn=size(node,1);
ndof=2*nn;        %Every node has 2 dofs.
ne=size(conn,1);         %this defines the size of elemental matrix
figure
for e=1:ne
    n1=conn(e,1);        %  first node for the element
    n2=conn(e,2);        % second node for the element

    x1=node(n1,1);       % X1 co-ordinate of the first node
    y1=node(n1,2);       % Y1 co-ordinate of the first node
    x2=node(n2,1);       % X2 co-ordinate of the second node
```

Figure 2.21 shows the trusss as drawn by the program.

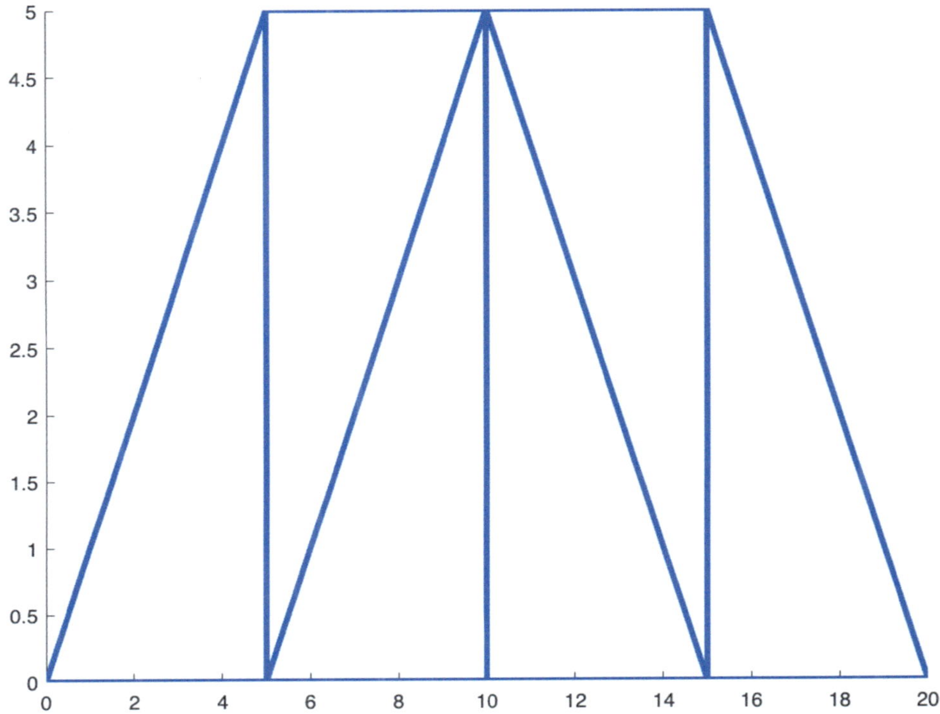

Fig. 2.21 Bridge-like Truss model

```
        y2=node(n2,2);        % Y2 co-ordinate of the second node

X = [x1,x2];
Y = [y1,y2];

line(X,Y,'linewidth',3);
end

%The force vector is defined using the data available from the
problem.
%First the vector is initialized to all zero.  Then the non-zero
values are
%added
F=zeros(2*nn,1);
F(12) = -1000E3;
F(14)= -1000E3;
F(16)= -1000E3;
```

```matlab
% Initalization of the Stiffness mmatrix and the displacemment vector
K=zeros(ndof,ndof);      %Defines size of the global stiffness matrix
d=zeros(ndof,1);         %Defines size of the displacemetn matrix
% This loop is executed for all the elements. The loop calculates
%element lengths and the cosine and sine of its angular orientation
% then it  creates  the stiffness mmatrix for each elemment in the
variable
% ke.  Finally the global stiffness matrix is assembled.
 for e=1:ne
     n1=conn(e,1);       %  first node for the element
     n2=conn(e,2);       % second node for the element

     x1=node(n1,1);      % X1 co-ordinate of the first node
     y1=node(n1,2);      % Y1 co-ordinate of the first node
     x2=node(n2,1);      % X2 co-ordinate of the second node
     y2=node(n2,2);      % Y2 co-ordinate of the second node

     L=sqrt((((x2-x1))^2+((y2-y1)^2))); % Length of the truss element

     c=(x2-x1)/L;     %cosine of element orientation
     s=(y2-y1)/L;     %sine of element orientation

     % ke is the element matrix and all its members are populated
     ke=((A*E)/L)*[c*c,s*c,-c*c,-s*c;s*c,s*s,-s*c,-s*s;-c*c,-
s*c,c*c,s*c;-s*c,-s*s,s*c s*s];

     % Here, we globalize the global matrix according to the
respective node
     % and respective displacement
     sctr=[2*n1-1 2*n1 2*n2-1 2*n2];
     K(sctr,sctr)=K(sctr,sctr)+ke;
 end


% We have got our global matrix. Now we solve the matrices to get the
% displacements at each node
d(isol)=K(isol,isol)\F(isol);
fprintf('\n----------Nodal Displacements----------');
```

----------Nodal Displacements----------

```matlab
fprintf('\nNo.  X-Direction      Y-Direction');
```

No. X-Direction   Y-Direction

```matlab
for i=1:nn
    fprintf('\n%5d  %8.3e   %8.3e',i,d(2*i-1),d(2*i));
end
```

```
 1  0.000e+00  0.000e+00
 2  0.000e+00  0.000e+00
 3  2.500e-02  -9.571e-02
 4  7.804e-18  -1.484e-01
 5  -2.500e-02  -9.571e-02
 6  -4.167e-03  -1.207e-01
 7  8.563e-18  -1.651e-01
 8  4.167e-03  -1.207e-01
```

```matlab
%Once all the displacements are found the reactions at each node are
%calculated

f=K*d;
fprintf('\n----------Reactions----------');
```

----------Reactions----------

```matlab
fprintf('\nNo.  X-Direction       Y-Direction');
```

No. X-Direction   Y-Direction

```matlab
for i=1:nn
    fprintf('\n%5d  %8.3e   %8.3e',i,f(2*i-1),f(2*i));
end
```

```
 1  1.750e+06  1.500e+06
 2  -1.750e+06  1.500e+06
 3  7.268e-10  -9.533e-10
 4  -6.746e-10  -7.224e-10
 5  1.886e-10  2.740e-10
 6  2.568e-10  -1.000e+06
 7  -2.789e-10  -1.000e+06
 8  2.089e-10  -1.000e+06
```

```matlab
for e=1:ne

    n1=conn(e,1);
    n2=conn(e,2);

    x1=node(n1,1);
    y1=node(n1,2);
```

```
    x2=node(n2,1);
    y2=node(n2,2);
    x1new = node(n1,1)+d(2*n1-1);
    y1new = node(n1,2)+d(2*n1);
    x2new = node(n2,1)+d(2*n2-1);
    y2new = node(n2,2)+d(2*n2);

    X = [x1new,x2new];
    Y = [y1new,y2new];
line(X,Y,'linewidth',3,'color',[1 0 1]);
end


% Now we print the strains and stresses
fprintf('\n----------Elemental strain & stress----------');
```

----------Elemental strain & stress----------

Figure 2.22 shows the original and the displaced structure on the same figure.

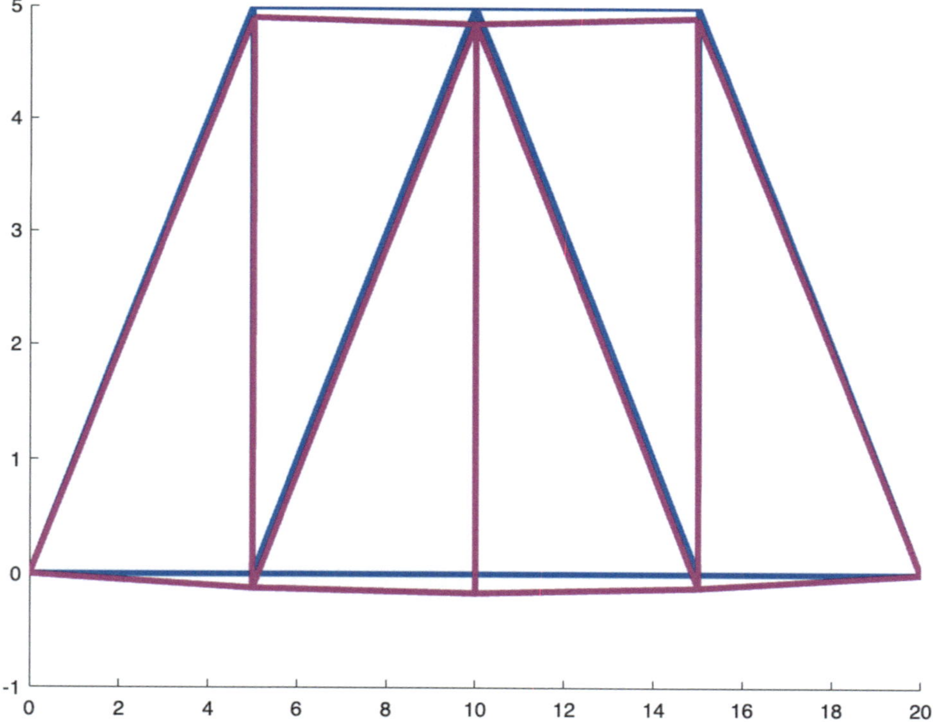

Fig. 2.22 Original and displaced truss

```matlab
fprintf('\nNo.  Strain  Stress');
for e=1:ne

    n1=conn(e,1);
    n2=conn(e,2);

    x1=node(n1,1);
    y1=node(n1,2);
    x2=node(n2,1);
    y2=node(n2,2);

    L=sqrt(((x2-x1))^2+((y2-y1)^2));
    c=(x2-x1)/L;
    s=(y2-y1)/L;

    % By following the standard calculations the strains and stresses are
    % calculated
    B=(1/L)*[-c -s c s];

    % Again it is calculated by selecting the respective values from the
    % globalized stiffness matrix
    sctr=[2*n1-1 2*n1 2*n2-1 2*n2];
    Strain=B*d(sctr);
    Stress=E*Strain;

    fprintf('\n%5d  %8.3e   %8.3e',e,Strain,Stress);
end
```

No. Strain Stress

```
1 -7.071e-03  -1.414e+09
2 -8.333e-04  -1.667e+08
3 5.000e-03  1.000e+09
4 -2.357e-03  -4.714e+08
5 8.333e-04  1.667e+08
6 -5.000e-03  -1.000e+09
7 3.333e-03  6.667e+08
8 8.333e-04  1.667e+08
9 -2.357e-03  -4.714e+08
10 -5.000e-03  -1.000e+09
11 5.000e-03  1.000e+09
12 -8.333e-04  -1.667e+08
13 -7.071e-03  -1.414e+09
```

2.4 Summary

In this chapter we explored the different uses of discrete finite elements. Discrete finite elements originated from the matrix method of solving structural problems. This method is used for static structures where members are naturally disconnected such as trusses, or spring—mass problems. We have presented examples of one-dimensional static problems, one-dimensional dynamic problems and two-dimensional truss problems.

Finite Elements for the Continuum, Heat Transfer Problems

3

In the previous chapter we reviewed the FEA approach used for discrete problems where the system naturally offers discrete components that are only attached at their end points. Discrete components in structures interact with each other, and the rules governing these interactions are the force-displacement relationships. We have learned how the development of the element equations result in the formulations of discrete finite elements.

In this chapter we will briefly discuss the very basic approach to finite element formulation for the continuum. Most real-life problems have to do with the continuum. Be it the transfer of energy or heat, flow of fluid, induction of a magnetic field or formation of high stress in a loaded component, most of the time we are dealing with a continuum where the variable that is of interest for us is varying across the entire domain. We will primarily use heat transfer as a phenomenon to go through this introduction of Continuum FEA. In the next chapter we will cover stress analysis problems.

3.1 The FEA Process

Before we move forward anymore with this discussion, we will define a few terms that we will use repeatedly. Domain is the term used to represent the area or volume that we are concerned with in a given problem. Boundary is the outer boundary of this domain. Figure 3.1 shows a schematic of this.

Within a domain the physical phenomena that are under consideration are governed by basic laws of physics. These laws of nature lead to mathematical expressions and these are usually in the form of partial (or sometimes ordinary) differential equations. The equations are typically partial since the quantity that is of interest could vary across the domain in all directions as well as with time (for time dependent problems).

One of the most common equations that shows up in many types of applications is the second order partial differential equation. An example of such an equation is the following:

© The Author(s), under exclusive license to Springer Nature Switzerland AG 2023 71
S. Das, *An Introduction to Finite Element Analysis Using Matlab Tools*,
Synthesis Lectures on Mechanical Engineering,
https://doi.org/10.1007/978-3-031-17540-4_3

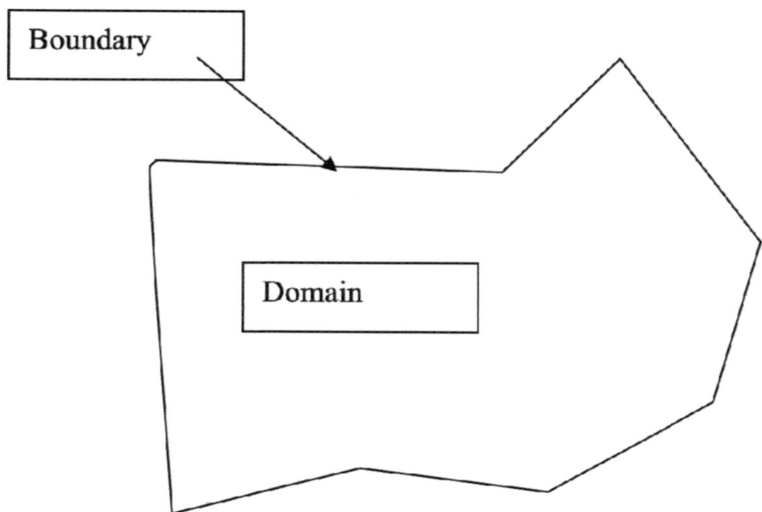

Fig. 3.1   Schematic of a 2D domain and boundary

$$k\left[\frac{\partial^2 U}{\partial x^2} + \frac{\partial^2 U}{\partial y^2}\right] = q \qquad (3.1)$$

Within this basic equation U is called the primary variable and x and y are the co-ordinates of any particular location. q is called the forcing function and k is a material property. This basic equation is the governing relationship for many practical engineering problems. The table below provides a partial list.

Typical applications of this equation are listed in the table

Field of application	Primary variable	Material constant	Forcing function
Steady state heat transfer	Temperature	Conductivity	Heat source
Irrotational flow of an ideal fluid	Stream function	Density	Mass function
Ground water flow	Piezometric head	Permeability	Recharge
Torsion of a constant cross-section	Stress function	$1/G$, G = shear stress	Angle of twist
Electrostatics	Scalar potential	Dielectric constant	Charge density
Magnetostatics	Magnetic potential	Magnetic permeability	Charge density
Transverse deflection of electric beams	Transverse deflection	Tension in membrane	Transversely distributed load

We will establish the process of solving these types of equations using the finite element approach. In our discussion we will not spend much time on the physical principles which were used to arrive at these equations or where these equations came from. We will instead assume that the equation has been given to us and we are responsible for developing suitable solution methods. Also, for now we will work on steady-state problems, i.e. the solution is not time dependent.

Before we move forward with establishing the process it is worth mentioning that along with the differential equations we need boundary conditions in order to solve these equations and typical boundary conditions that could be applicable for these types of equations are:

$$U, given$$
$$\frac{\partial U}{\partial n}, given$$
$$aU + b\frac{\partial U}{\partial n}, given \tag{3.2}$$

where n is the outer normal to the surface.

In order to understand how the process works for continuum FEA let us consider a second order differential equation in 1-D. We choose 1-D so that we can follow along with the mathematics relatively easily and would not be bogged down by the complexities in geometry.

So, the generic differential equation is written as:

$$a\frac{d^2u}{dx^2} + u = f \tag{3.3}$$

Let the domain on which this equation is valid extend from x = 0 to x = L.

To solve this equation using the finite element approach we go through the following steps:

- multiply the equation with a test function w(x). At this point we are not specifying what the w(x) function is except that it is a function of x.

$$aw\frac{d^2u}{dx^2} + uw = fw \tag{3.4}$$

- Now we integrate this equation over the whole domain:

$$\int aw\frac{d^2u}{dx^2}dx + \int uwdx = \int fwdx \tag{3.5}$$

After applying the limits of integration, the equation looks like:

$$\int_0^L aw\frac{d^2u}{dx^2}dx + \int_0^L uwdx = \int_0^L fwdx \tag{3.6}$$

- If we consider the three terms in the above equation to be terms I, II and III, let us consider the first term by itself. By applying the rules of integration by parts to Term I we get

$$\int_0^L aw\frac{d^2u}{dx^2}dx = \left[aw\frac{du}{dx}\right]_0^L - \int_0^L a\frac{dw}{dx}\frac{du}{dx}dx \tag{3.7}$$

- By inserting this in the original integral equation we get:

$$\int_0^L uwdx + \left[aw\frac{du}{dx}\right]_0^L - \int_0^L a\frac{dw}{dx}\frac{du}{dx}dx = \int_0^L fwdx \tag{3.8}$$

This form where the second order term is replaced by a product of first order terms and the boundary terms is called the weak form. And the weak form is an integral form of the original differential equation.

Instead of solving the differential equation, in the FE approach we will try to solve the integral equation that is a representation of the original differential equation. Figure 3.2 shows the domain; let the domain be divided into a large number of elements. In the figure, one such element is shown whose end points are generic locations x1 and x2 (Fig. 3.3).

Using the many elements that this domain has been divided into we can re-write the integral (which is over the whole domain) as a summation of integrals over each element. So, if there are N elements the equation may be re-written as:

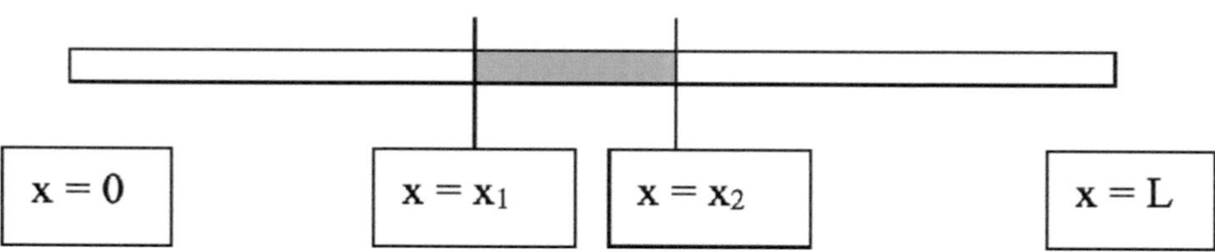

Fig. 3.2 The 1-D domain and a generic element in the domain

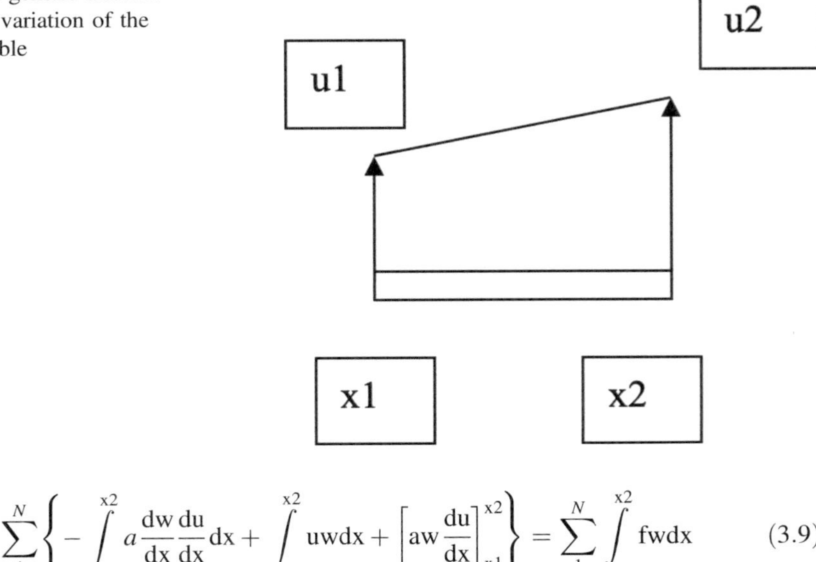

Fig. 3.3 The generic element and assumed variation of the primary variable

$$\sum_{1}^{N}\left\{-\int_{x1}^{x2} a\frac{dw}{dx}\frac{du}{dx}dx + \int_{x1}^{x2} uwdx + \left[aw\frac{du}{dx}\right]_{x1}^{x2}\right\} = \sum_{1}^{N}\int_{x1}^{x2} fwdx \qquad (3.9)$$

We will now attempt to integrate all these functions and assemble them for all the elements so that the above equation may be satisfied.

As a first step we assume that u, the primary variable varies linearly between x1 and x2 and its values are u1 and u2 at these two end points, respectively. So over the element the primary variable may be expressed as a linear function of x as:

$$u = ax + b \qquad (3.10)$$

By substituting the end point values of u for the two end nodes the two variables a and b can be computed and the equation for u may be written as:

$$u = \frac{(x2 - x)}{(x2 - x1)}u1 + \frac{(x - x1)}{(x2 - x1)}u2 \qquad (3.11)$$

This way, u is written as a linear function but is related to the values at the end points of the element, u1 and u2. The multipliers are called shape functions, N1 and N2 where:

$$N1(x) = \frac{(x2 - x)}{(x2 - x1)} \qquad (3.12)$$

$$N2(x) = \frac{(x - x1)}{(x2 - x1)} \qquad (3.13)$$

Properties of shape functions:

If you consider the shape functions carefully there are two properties of shape functions that come in view.

N1 + N2 = 1 (at all times)
N1(x = x1) = 1, N2(x = x1) = 0; N1(x = x2) = 0, N2(x = x2) = 1

Visually the properties can be represented as shown in Fig. 3.4.

In considering higher order elements we need to remember that these properties need to be satisfied.

Now that we have approximated the behavior of u on any one element as a linear function we can try to work with the terms in the integral equation and try and solve it. As we saw before the equation is:

$$\sum_{1}^{N}\left\{ -\int_{x1}^{x2} a\,\frac{dw}{dx}\frac{du}{dx}\,dx + \int_{x1}^{x2} uw\,dx + \left[aw\frac{du}{dx} \right]_{x1}^{x2} \right\} = \sum_{1}^{N}\int_{x1}^{x2} fw\,dx \qquad (3.14)$$

Consider the above equation and for the moment we will not worry about the summation sign in front. Since we are approximating the u as a linear function written in terms of the shape function, we will substitute that. Earlier, w was only defined as a test function that is a function of x. So, we will now say that w is first taken as N1 and then as N2 (which are both functions of x) and we write the above equation for an element twice. We get,

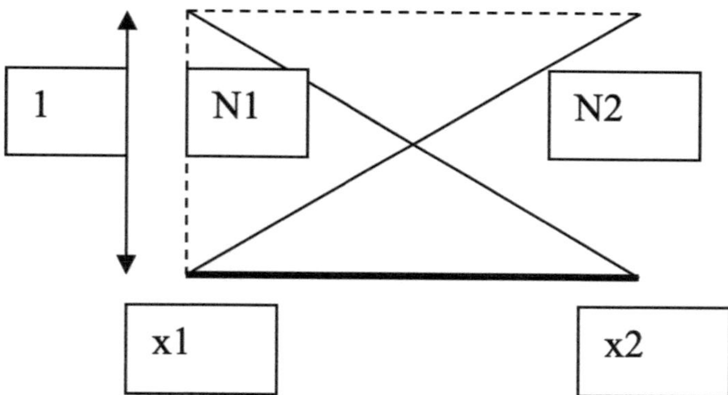

Fig. 3.4 Properties of linear shape functions

$$-\int_{x1}^{x2} a\frac{dN1}{dx}\left(u1\frac{dN1}{dx}+u2\frac{dN2}{dx}\right)dx + \int_{x1}^{x2}(u1N1+u2N2)\,N1dx$$

$$+\left[aN1\left(u1\frac{dN1}{dx}+u2\frac{dN2}{dx}\right)\right]_{x1}^{x2} \tag{3.15}$$

and

$$\int_{x1}^{x2}(fN1)dx \tag{3.16}$$

for the left-hand side and the right-hand side respectively. And

$$-\int_{x1}^{x2} a\frac{dN2}{dx}\left(u1\frac{dN1}{dx}+u2\frac{dN2}{dx}\right)dx + \int_{x1}^{x2}(u1N1+u2N2)N2dx$$

$$+\left[aN2\left(u1\frac{dN1}{dx}+u2\frac{dN2}{dx}\right)\right]_{x1}^{x2} \tag{3.17}$$

and

$$\int_{x1}^{x2}(fN2)dx \tag{3.18}$$

for the left-hand side and right-hand side of the second equation respectively.

These equations result in a number of computations of integral over a single element.

In order to understand the nature of each term in the equations let us consider each term separately. So, the first integral in the two equations are as follows:

$$-\int_{x1}^{x2} a\frac{dN2}{dx}\left(u1\frac{dN1}{dx}+u2\frac{dN2}{dx}\right)dx$$

$$=-\int_{x1}^{x2}\left(au1\frac{dN2}{dx}\frac{dN1}{dx}+au2\frac{dN2}{dx}\frac{dN2}{dx}\right)dx \tag{3.19}$$

$$=-au1\int_{x1}^{x2}\frac{dN2}{dx}\frac{dN1}{dx}dx - au2\int_{x1}^{x2}\frac{dN2}{dx}\frac{dN2}{dx}dx$$

$$-\int_{x1}^{x2} a\frac{dN2}{dx}\left(u1\frac{dN1}{dx}+u2\frac{dN2}{dx}\right)dx = -\int_{x1}^{x2}\left(au1\frac{dN2}{dx}\frac{dN1}{dx}+au2\frac{dN2}{dx}\frac{dN2}{dx}\right)dx$$

$$= -au1\int_{x1}^{x2}\frac{dN2}{dx}\frac{dN1}{dx}dx - au2\int_{x1}^{x2}\frac{dN2}{dx}\frac{dN2}{dx}dx \tag{3.20}$$

The multiplier a comes from the original differential equation and is assumed to be a constant. u1 and u2 are specific values of the primary variable at the two nodes of the element and are not varying with x (or across the element). So those quantities can move outside the integral sign. The terms left inside the integral are the first derivatives of the shape functions. Let's look at what those expressions are:

$$N1(x) = \frac{(x2-x)}{(x2-x1)} \tag{3.21}$$

$$N2(x) = \frac{(x-x1)}{(x2-x1)} \tag{3.22}$$

So

$$\frac{dN1}{dx} = \frac{(-1)}{(x2-x1)} = \frac{-1}{L} \tag{3.23}$$

$$\frac{dN2}{dx} = \frac{(1)}{(x2-x1)} = \frac{1}{L} \tag{3.24}$$

where L is the length of the element.

$$-au1\int_{x1}^{x2}\frac{dN1}{dx}\frac{dN1}{dx}dx - au2\int_{x1}^{x2}\frac{dN1}{dx}\frac{dN2}{dx}dx$$

$$= -au1\left(\frac{(x2-x1)}{L^2}\right) - au2\left(\frac{-(x2-x1)}{L^2}\right) \tag{3.25}$$

$$= -au1\left(\frac{1}{L}\right) + au2\left(\frac{1}{L}\right)$$

$$-au1 \int_{x1}^{x2} \frac{dN2}{dx}\frac{dN1}{dx}dx - au2 \int_{x1}^{x2} \frac{dN2}{dx}\frac{dN2}{dx}dx$$

$$= -au1\left(\frac{-(x2-x1)}{L^2}\right) - au2\left(\frac{(x2-x1)}{L^2}\right) \tag{3.26}$$

$$= au1\left(\frac{1}{L}\right) - au2\left(\frac{1}{L}\right)$$

So in matrix form the two results from (3.25) and (3.26) can be written as:

$$-\frac{a}{L}\begin{bmatrix} 1 & -1 \\ -1 & 1 \end{bmatrix}\begin{bmatrix} u1 \\ u2 \end{bmatrix} \tag{3.27}$$

Now let us consider the second term in the original two relations (3.15 and 3.17):

$$\int_{x1}^{x2} (u1N1 + u2N2)N1dx = u1 \int_{x1}^{x2} N1N1dx + u2 \int_{x1}^{x2} N1N2dx \tag{3.28}$$

$$\int_{x1}^{x2} (u1N1 + u2N2)N2dx = u1 \int_{x1}^{x2} N1N2dx + u2 \int_{x1}^{x2} N2N2dx \tag{3.29}$$

If we compute these integrals we will get:

$$u1 \int_{x1}^{x2} N1N1dx = \frac{u1}{L^2} \int_{x1}^{x2} (x2-x)^2 dx = \frac{u1}{L^2}\left[x2^2 x - 2x^2\frac{x2}{2} + \frac{x^3}{3}\right]_{x1}^{x2}$$

$$= \frac{u1}{L^2}\left[\frac{x2^3}{3} - \left(x2^2 x1 - x1^2 x2 + \frac{x1^3}{3}\right)\right] = \frac{u1}{L^2}\frac{(x2-x1)^3}{3} \tag{3.30}$$

So, the first term in Eq. 3.28 is

$$\frac{u1 L}{3} \tag{3.31}$$

Similarly, by integrating all the terms and arranging them in a matrix for we get:

$$\frac{L}{6}\begin{bmatrix} 2 & 1 \\ 1 & 2 \end{bmatrix}\begin{bmatrix} u1 \\ u2 \end{bmatrix} \tag{3.32}$$

And the terms from the right-hand side (expressions 3.16 and 3.18) have a function f in them. f comes from the original differential equation and can be a constant or a function of x in a given problem. There are several ways to handle this. Once we know the actual function, we can carry out the integral. Or we could generically carry out the integral by approximating f to be a constant value over given element (or it could be assumed to be linear or any other form on the element). If we assume f to be a constant on a given element (and a different constant on another element), we can represent f symbolically as f^e. And the terms can be written as:

$$\int_{x1}^{x2} (fN1)dx = f^e \int_{x1}^{x2} N1dx = f^e \int_{x1}^{x2} \frac{(x2-x)}{(x2-x1)}dx$$
$$= \frac{f^e}{L}\left[x2x - \frac{x^2}{2}\right]_{x1}^{x2} = \frac{f^e}{2L}(x2-x1)^2 = \frac{f^e}{2}L \tag{3.33}$$

$$\int_{x1}^{x2} (fN2)dx = f^e \int_{x1}^{x2} N2dx = f^e \int_{x1}^{x2} \frac{(x-x1)}{(x2-x1)}dx$$
$$= \frac{f^e}{L}\left[\frac{x^2}{2} - x1x\right]_{x1}^{x2} = \frac{f^e}{2L}(x2-x1)^2 = \frac{f^e}{2}L \tag{3.34}$$

So, in matrix form the right-hand side can be written as:

$$\frac{f^e L}{2}\begin{bmatrix} 1 \\ 1 \end{bmatrix} \tag{3.35}$$

This leaves two other terms in the two Eqs. (3.15 and 3.17). They are the boundary terms (not integrals)

$$-\left[aN1\left(u1\frac{dN1}{dx} + u2\frac{dN2}{dx}\right)\right]_{x1}^{x2} = 0 + a(1)\frac{dU}{dx}(x1) \tag{3.36}$$

$$-\left[aN2\left(u1\frac{dN1}{dx} + u2\frac{dN2}{dx}\right)\right]_{x1}^{x2} = -a(1)\frac{dU}{dx}(x2) + 0 \tag{3.37}$$

The results become simpler because N1 is 0 at x2 and N2 is 0 at x1.
Let us look at some examples how this is implemented.

Example 3.1 Consider a fairly simple example:

A bar of length 10 units. The temperature at the left end is 100 and at the right end is 50. This is divided into five elements of length 2 units each (Fig. 3.5).

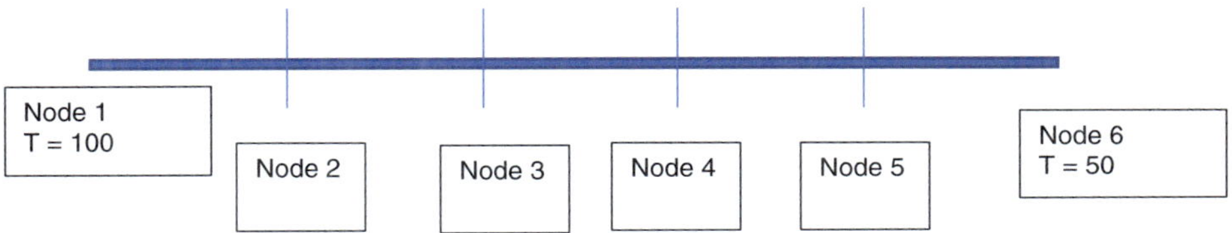

Fig. 3.5 Schematic for Example 3.1

The governing differential equation is:

$$k\frac{d^2T}{dx^2} = 0;\tag{3.38}$$

with boundary conditions, @ x = 0 T = 100, @ x = 10, T = 50.

Since this is a very simple equation, we can write its analytical solution by integrating:

$$\frac{dT}{dx} = C1;$$
$$T = C1x + C2\tag{3.39}$$

Applying the boundary conditions we get:

$$100 = C1(0) + C2; \quad C2 = 100$$
$$50 = C1\,(10) + 100; \quad C1 = -50/10 = -5\tag{3.40}$$
$$So, \quad T = 100 - 5x$$

By substituting the value of x in this equation we can calculate the temperature anywhere that varies linearly across the length of the bar.

Now considering the FEA solution,

The weak form is for one element:

$$\left[kw\frac{dT}{dx}\right]_{x1}^{x2} - \int_{x1}^{x2} k\frac{dw}{dx}\frac{dT}{dx}dx = 0\tag{3.41}$$

Using what we derived before, the integral may be written in a matrix form as

$$-\frac{k}{L}\begin{bmatrix} 1 & -1 \\ -1 & 1 \end{bmatrix}\tag{3.42}$$

Assembling this matrix for 5 elements we get:

$$-\frac{k}{L}\begin{bmatrix} 1 & -1 & 0 & 0 & 0 & 0 \\ -1 & 2 & -1 & 0 & 0 & 0 \\ 0 & -1 & 2 & -1 & 0 & 0 \\ 0 & 0 & -1 & 2 & -1 & 0 \\ 0 & 0 & 0 & -1 & 2 & -1 \\ 0 & 0 & 0 & 0 & -1 & 1 \end{bmatrix}\begin{bmatrix} T_1 \\ T_2 \\ T_3 \\ T_4 \\ T_5 \\ T_6 \end{bmatrix} = -k\begin{bmatrix} -\frac{dT}{dx}_{x1} \\ 0 \\ 0 \\ 0 \\ 0 \\ \frac{dT}{dx}_{x2} \end{bmatrix} \tag{3.43}$$

Since T1 = 100 and T6 = 50 we apply that condition at take the middle four equations:

$$\begin{bmatrix} 2 & -1 & 0 & 0 \\ -1 & 2 & -1 & 0 \\ 0 & -1 & 2 & -1 \\ 0 & 0 & -1 & 2 \end{bmatrix}\begin{bmatrix} T_2 \\ T_3 \\ T_4 \\ T_5 \end{bmatrix} + \begin{bmatrix} -100 \\ 0 \\ 0 \\ -50 \end{bmatrix} = L\begin{bmatrix} 0 \\ 0 \\ 0 \\ 0 \end{bmatrix} \tag{3.44}$$

$$\begin{bmatrix} 2 & -1 & 0 & 0 \\ -1 & 2 & -1 & 0 \\ 0 & -1 & 2 & -1 \\ 0 & 0 & -1 & 2 \end{bmatrix}\begin{bmatrix} T_2 \\ T_3 \\ T_4 \\ T_5 \end{bmatrix} = \begin{bmatrix} 100 \\ 0 \\ 0 \\ 50 \end{bmatrix} \tag{3.45}$$

```
K = [2 -1 0 0;-1 2 -1 0;0 -1 2 -1;0 0 -1 2]

K = 4×4
     2    -1     0     0
    -1     2    -1     0
     0    -1     2    -1
     0     0    -1     2

b = [100;0; 0; 50]

b = 4×1
   100
     0
     0
    50

K\b

ans = 4×1
   90.0000
   80.0000
   70.0000
   60.0000
```

So the temperatures at the six nodes are 100, 90, 80, 70, 60, 50.

Example 3.2 Consider the problem of heat transfer through a fin. Fins are used to increase heat transfer (loss) from hot tube or surfaces by increasing the effective surface area from which heat may be removed (Fig. 3.6).

The differential equation for heat transfer through the fin may be written as:

$$\frac{d}{dx}\left(kA\frac{dT}{dx}\right) = -hP(T - T_\infty) \tag{3.46}$$

By moving terms around we get:

$$\frac{d}{dx}\left(kA\frac{dT}{dx}\right) - hP(T) = hP(T_\infty) \tag{3.47}$$

In the equation k is the thermal conductivity, A is the cross-section area of the fin, P is the perimeter and h are the film heat transfer coefficient, T_∞ is the ambient temperature.

Typical boundary conditions in this situation are a constant temperature at the base of the fin, i.e. T(0) will be given as the wall temperature and convective boundary condition at the other end.

$$kA\frac{dT}{dx} = -hA(T - T_\infty)\text{at } x = L \tag{3.48}$$

In this example let T(0) = 100.

And k = 237 W/m K, h = 30 W/m^2 K, L = 20 cm, W = 3 m, T_∞ = 25 C and thickness of fin t = 0.003 m. So P = 6.006 m, A = 3 * 0.003 = 0.009 m^2, kA = 237 * 0.009 = 2.133 W m/K, hP = 180.18 W/m K, hpT_∞ = 4504.5 W/m, hA = 30 * 0.009 = 0.27 W/K.

Let us say we have divided the length into four elements so the nodes are at 0, 0.05, 0.1, 0.15, 0.2.

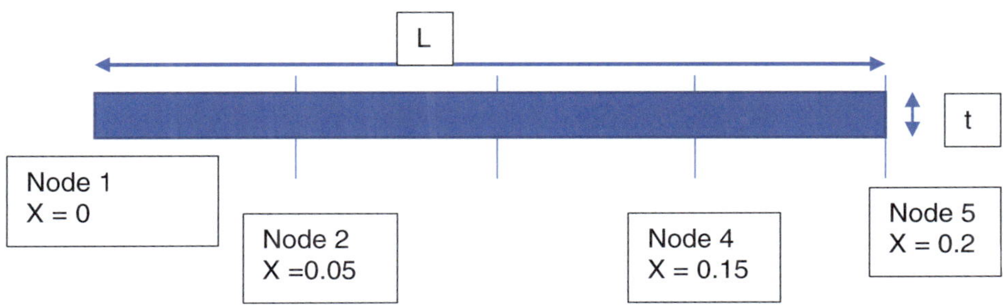

Fig. 3.6 Schematic for Example 3.2 the fin

Using the approach discussed before the first term in the differential equation results is a 2×2 matrix and some boundary terms, the second term results in a 2×2 matrix and the term on the right-hand side results in a forcing function. So, the matrix equation based on the derived results may be written as (for each element):

$$a\frac{d^2u}{dx^2} + u = f \tag{3.49}$$

If we compare the given equation with the generic equation used in the derivation (3.49) we can go through the entire process and write the matrix equation as (the inequality sign is used since the equality does not apply at the element level but will be true once the equations are assembled):

$$-\frac{kA}{L}\begin{bmatrix} 1 & -1 \\ -1 & 1 \end{bmatrix}\begin{bmatrix} T1 \\ T2 \end{bmatrix} - \frac{hPL}{6}\begin{bmatrix} 2 & 1 \\ 1 & 2 \end{bmatrix}\begin{bmatrix} T1 \\ T2 \end{bmatrix} + \begin{bmatrix} \left(-kA\frac{dT}{dx}\right)_{x1} \\ \left(kA\frac{dT}{dx}\right)_{x2} \end{bmatrix} \neq \frac{-hPLT_\infty}{2}\begin{bmatrix} 1 \\ 1 \end{bmatrix} \tag{3.50}$$

The inequality sign means that this left-hand side is not equal to this right-hand side. They will be equal once all the elements for the entire model are assembled.

Element#1

$$-\frac{2.133}{0.05}\begin{bmatrix} 1 & -1 \\ -1 & 1 \end{bmatrix}\begin{bmatrix} T1 \\ T2 \end{bmatrix} - \frac{180.18 * 0.05}{6}\begin{bmatrix} 2 & 1 \\ 1 & 2 \end{bmatrix}\begin{bmatrix} T1 \\ T2 \end{bmatrix}$$
$$\neq -\begin{bmatrix} \left(-kA\frac{dT}{dx}\right)_{x1} \\ \left(kA\frac{dT}{dx}\right)_{x2} \end{bmatrix} + \frac{-(4504.5) * (0.05)}{2}\begin{bmatrix} 1 \\ 1 \end{bmatrix} \tag{3.51}$$

$$\begin{bmatrix} 45.663 & -41.1585 \\ -41.1585 & 45.663 \end{bmatrix}\begin{bmatrix} T1 \\ T2 \end{bmatrix} \neq \begin{bmatrix} \left(-kA\frac{dT}{dx}\right)_{x1} \\ \left(kA\frac{dT}{dx}\right)_{x2} \end{bmatrix} + 112.613\begin{bmatrix} 1 \\ 1 \end{bmatrix} \tag{3.52}$$

Element #2

$$\begin{bmatrix} 45.663 & -41.1585 \\ -41.1585 & 45.663 \end{bmatrix}\begin{bmatrix} T2 \\ T3 \end{bmatrix} \neq \begin{bmatrix} \left(-kA\frac{dT}{dx}\right)_{x2} \\ \left(kA\frac{dT}{dx}\right)_{x3} \end{bmatrix} + 112.613\begin{bmatrix} 1 \\ 1 \end{bmatrix} \tag{3.53}$$

Element #3

$$\begin{bmatrix} 45.663 & -41.1585 \\ -41.1585 & 45.663 \end{bmatrix} \begin{bmatrix} T3 \\ T4 \end{bmatrix} \neq \begin{bmatrix} \left(-kA\frac{dT}{dx}\right)_{x3} \\ \left(kA\frac{dT}{dx}\right)_{x4} \end{bmatrix} + 112.613 \begin{bmatrix} 1 \\ 1 \end{bmatrix} \qquad (3.54)$$

Element #4

$$\begin{bmatrix} 45.663 & -41.1585 \\ -41.1585 & 45.663 \end{bmatrix} \begin{bmatrix} T4 \\ T5 \end{bmatrix} \neq \begin{bmatrix} \left(-kA\frac{dT}{dx}\right)_{x4} \\ \left(kA\frac{dT}{dx}\right)_{x5} \end{bmatrix} + 112.613 \begin{bmatrix} 1 \\ 1 \end{bmatrix} \qquad (3.55)$$

If we assemble these equations we get:

$$\begin{bmatrix} 45.663 & -41.1585 & 0 & 0 & 0 \\ -41.1585 & 91.326 & -41.1585 & 0 & 0 \\ 0 & -41.1585 & 91.326 & -41.1585 & 0 \\ 0 & 0 & -41.1585 & 91.326 & -41.1585 \\ 0 & 0 & 0 & -41.1585 & 45.663 \end{bmatrix} \begin{bmatrix} T1 \\ T2 \\ T3 \\ T4 \\ T5 \end{bmatrix}$$
$$= \begin{bmatrix} \left(-kA\frac{dT}{dx}\right)_{x1} \\ 0 \\ 0 \\ 0 \\ \left(kA\frac{dT}{dx}\right)_{x5} \end{bmatrix} + \begin{bmatrix} 112.613 \\ 225.225 \\ 225.225 \\ 225.225 \\ 112.613 \end{bmatrix} \qquad (3.56)$$

$$\begin{bmatrix} 45.663 & -41.1585 & 0 & 0 & 0 \\ -41.1585 & 91.326 & -41.1585 & 0 & 0 \\ 0 & -41.1585 & 91.326 & -41.1585 & 0 \\ 0 & 0 & -41.1585 & 91.326 & -41.1585 \\ 0 & 0 & 0 & -41.1585 & 45.663 \end{bmatrix} \begin{bmatrix} T1 \\ T2 \\ T3 \\ T4 \\ T5 \end{bmatrix}$$
$$= \begin{bmatrix} \left(-kA\frac{dT}{dx}\right)_{x1} \\ 0 \\ 0 \\ 0 \\ -0.27(T5 - T_\infty) \end{bmatrix} + \begin{bmatrix} 112.613 \\ 225.225 \\ 225.225 \\ 225.225 \\ 112.613 \end{bmatrix} \qquad (3.57)$$

$$
\begin{bmatrix}
45.663 & -41.1585 & 0 & 0 & 0 \\
-41.1585 & 91.326 & -41.1585 & 0 & 0 \\
0 & -41.1585 & 91.326 & -41.1585 & 0 \\
0 & 0 & -41.1585 & 91.326 & -41.1585 \\
0 & 0 & 0 & -41.1585 & 45.663
\end{bmatrix}
\begin{bmatrix} T1 \\ T2 \\ T3 \\ T4 \\ T5 \end{bmatrix}
$$

$$
= \begin{bmatrix}
\left(-kA\frac{dT}{dx}\right)_{x1} \\ 0 \\ 0 \\ 0 \\ -0.27(T5 - T_\infty)
\end{bmatrix}
+ \begin{bmatrix} 112.613 \\ 225.225 \\ 225.225 \\ 225.225 \\ 112.613 \end{bmatrix}
\tag{3.58}
$$

$$
\begin{bmatrix}
45.663 & -41.1585 & 0 & 0 & 0 \\
-41.1585 & 91.326 & -41.1585 & 0 & 0 \\
0 & -41.1585 & 91.326 & -41.1585 & 0 \\
0 & 0 & -41.1585 & 91.326 & -41.1585 \\
0 & 0 & 0 & -41.1585 & 45.663 + 0.27
\end{bmatrix}
\begin{bmatrix} T1 \\ T2 \\ T3 \\ T4 \\ T5 \end{bmatrix}
$$

$$
= \begin{bmatrix}
\left(-kA\frac{dT}{dx}\right)_{x1} \\ 0 \\ 0 \\ 0 \\ -0.27(-25)
\end{bmatrix}
+ \begin{bmatrix} 112.613 \\ 225.225 \\ 225.225 \\ 225.225 \\ 112.613 \end{bmatrix}
\tag{3.59}
$$

We have already incorporated the boundary condition at x = 20 cm. We need to incorporate the boundary conditions at x = 0. The BC at x = 0 is T = 100. If we apply that in the above equations (i.e. T1 = 100) and remove the first equation we get:

$$
\begin{bmatrix}
91.326 & -41.1585 & 0 & 0 \\
-41.1585 & 91.326 & -41.1585 & 0 \\
0 & -41.1585 & 91.326 & -41.1585 \\
0 & 0 & -41.1585 & 45.663 + 0.27
\end{bmatrix}
\begin{bmatrix} T2 \\ T3 \\ T4 \\ T5 \end{bmatrix}
$$

$$
= \begin{bmatrix}
225.225 + 41.1585 * 100 \\
225.225 \\
225.225 \\
112.613 + 0.27 * 25
\end{bmatrix}
\tag{3.60}
$$

Now we use Matlab commands to solve this equation.

```
A = [91.326 -41.1585 0 0;-41.1585 91.326 -41.1585 0;0 -41.1585 91.326 -
41.1585;0 0 -41.1585 45.663+0.27]

A = 4×4
    91.3260   -41.1585          0          0
   -41.1585    91.3260   -41.1585          0
          0   -41.1585    91.3260   -41.1585
          0          0   -41.1585    45.9330

B = [225.225+41.1585*100;225.225;225.225;112.613+0.27*25]

B = 4×1
10³ ×
     4.3411
     0.2252
     0.2252
     0.1194

T = A\B

T = 4×1
    73.8496
    58.3916
    50.2425
    47.6187
```

So, the temperatures are 100, 73.85, 58.39, 50.24, 47.62.

Let's plot them

```
x = [0 0.05 0.1 0.15 0.2]

x = 1×5
         0     0.0500     0.1000     0.1500     0.2000

T = [100, 73.85, 58.39, 50.24,47.62]

T = 1×5
   100.0000    73.8500    58.3900    50.2400    47.6200

plot(x,T,"LineWidth",3);
xlabel('distance');
ylabel('Temperature');
```

If we plot these we will get five-point values at the five nodes and the temperatures
(Fig. 3.7) will be interpolated linearly between the points since we assumed the temperature
to be linearly varying over each element. So, if we consider say the second element:

T2 (@ x = 0.05) = 73.8496.

T3 (@ x = 0.1) = 58.3916.

And if we want to compute the temperature at x = 0.075 we will have to do the following:

T (on second element) = N1 * T2 + N2 * T3

$$T = \frac{(x3 - x)}{(x3 - x2)} T2 + \frac{(x - x2)}{(x3 - x2)} T3$$

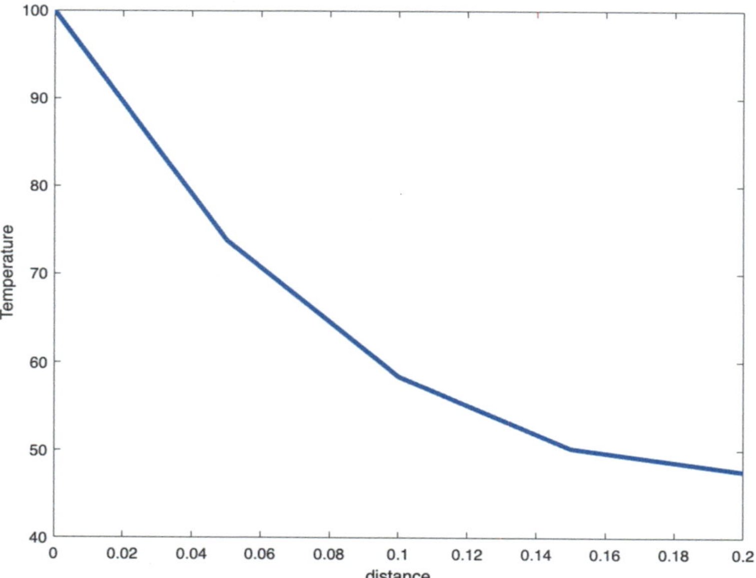

Fig. 3.7 Temperature distribution across the fine length

$$T = \frac{(0.1 - 0.075)}{(0.1 - 0.05)}\,73.8596 + \frac{(0.075 - 0.05)}{(0.1 - 0.05)}\,58.3916$$

```
Treq = (x(3)-0.075)*T(2)/(x(3)-x(2))+(0.075-x(2))*T(3)/(x(3)-x(2))
```
Treq = 66.1200

If we want to calculate the derivative of temperature which is the heat flux on an element we need to compute the derivative of T with respect to x, e.g. on the second element,

$$\frac{dT}{dx} = \frac{(-1)}{(x3 - x2)}\,T2 + \frac{(1)}{(x3 - x2)}\,T3$$

```
flux2 = (-1)*T(2)/(x(3)-x(2))+(1)*T(3)/(x(3)-x(2))
```

flux2 = −309.2000.

A similar calculation on a different but adjacent element (e.g. 1 and 3) will show a different number. This means that even though the temperature (primary variables) are continuous across the elements the derivatives (or secondary variables) are not.

```
flux3 = (-1)*T(3)/(x(4)-x(3))+(1)*T(4)/(x(4)-x(3))
```

flux3 = −163.0000.

Also, the primary variables are computed on the nodes but the secondary variables are calculated over the elements. So, if the primary variable is assumed to vary linearly across the elements the secondary will be constant across the elements. If the primary is chosen to vary quadratic across the elements, then the secondary will vary linearly across the element. The temperature plot also shows slope discontinuity to emphasize this factor as well.

In order to get better accuracy in the secondary variable a higher order element has to be chosen such as the quadratic or cubic.

Example 3.2 Using PDE Toolbox

This example uses the FEA tool within MATLAB (PDE toolbox) to model the same example as the one worked out before. The example is idealized as 1-D example in the previous analysis. But here it is modeled as a thin 2-D problem. The results however, will show that this problem essentially works as a 1-D problem.

The geometry is a long thin rod of dimension 0.2 m by 0.003 m. One end of the bar is held at a constant temperature and the bar experiences convective heat loss from all the other sides.

Steady-State Solution

Create a steady-state thermal model for solving a 2D problem.

```
thermalModelS = createpde('thermal','steadystate');% declaring the type of
analysis
g = decsg([3 4 0 .2 .2 0 0 0 0.003 0.003]');%defining thee geometry
```

Include the geometry in the model.

```
geometryFromEdges(thermalModelS,g);
```

Plot the geometry with the edge labels. (Fig. 3.8)

```
figure
pdegplot(thermalModelS,'EdgeLabels','on')
axis equal
```

The rod is composed of a material with these thermal properties. In a steady state heat transfer problem only the thermal conductivity is needed.

```
k = 237; % Thermal conductivity, W/(m*C)
```

For a steady-state analysis, specify the thermal conductivity of the material.

```
thermalProperties(thermalModelS,'ThermalConductivity',k);
```

Specify the internal heat source. In this particular example there is no internal heat source.

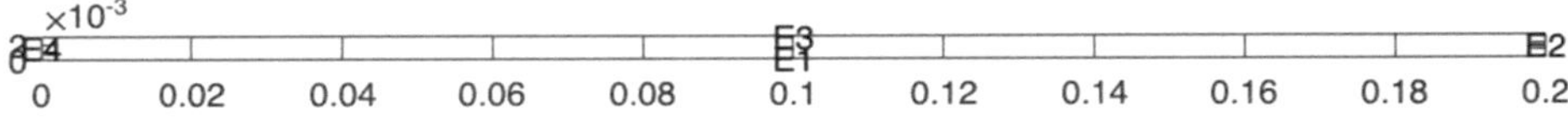

Fig. 3.8 Rod geometry

```
%internalHeatSource(thermalModelS,q);
```

Define the boundary conditions. Edge 4 is kept at a constant temperature $T = 100$ °C. All the other edges have convective boundary conditions.

```
thermalBC(thermalModelS,'Edge',4,'Temperature',100);
```

Specify the convection boundary condition on the outer boundary (edge 1, 2, 3). The surrounding temperature at the outer boundary is 25 °C, and the heat transfer coefficient is 30 W/(m^2 °C).

```
thermalBC(thermalModelS,'Edge',1,...
                        'ConvectionCoefficient',30,...
                        'AmbientTemperature',25);
thermalBC(thermalModelS,'Edge',2,...
                        'ConvectionCoefficient',30,...
                        'AmbientTemperature',25);
thermalBC(thermalModelS,'Edge',3,...
                        'ConvectionCoefficient',30,...
                        'AmbientTemperature',25);
% we are not using a heat flux BC in this case. Here is a sample command
% line if we were to use the known heat flux condition
%thermalBC(thermalModelS,'Edge',4,'HeatFlux',5000);
```

Generate the mesh (Fig. 3.9)

```
msh = generateMesh(thermalModelS,"GeometricOrder","linear","Hmax",0.005); %the
maximum element size defines control
% of the element size
figure
pdeplot(thermalModelS)
axis([0 0.2 -0.02 0.02])
```

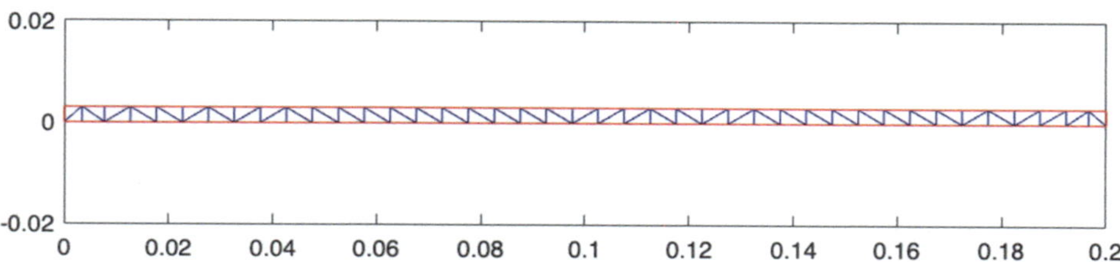

Fig. 3.9 Meshed geometry of rod

Solve the model and plot the result (Figs. 3.10 and 3.11).

```matlab
result = solve(thermalModelS);
T = result.Temperature;
flux = result.evaluateHeatFlux;
figure
pdeplot(thermalModelS,'XYData',T,'Contour','on','ColorBar','on','ColorMap','jet
')
axis equal
title 'Steady-State Temperature'

%plotting the temperature across the length of the bar
X = 0:0.01:0.2;
Y = zeros(size(X));

Tintrp = interpolateTemperature(result,X,Y);

plot(X,Tintrp,"LineWidth",3);
xlabel('distance');
ylabel('Temperature')
```

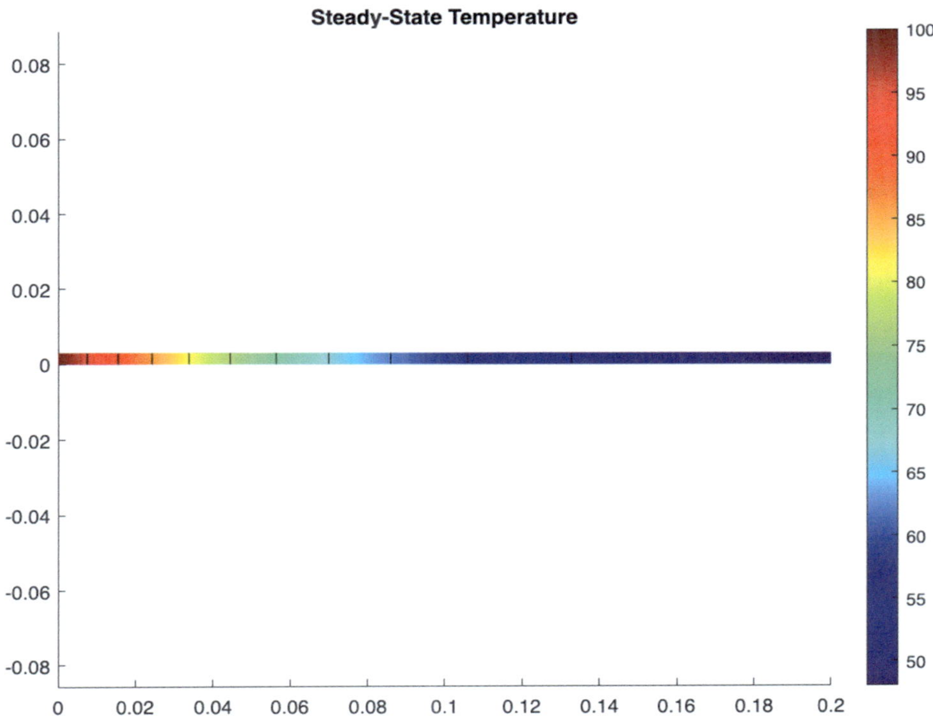

Fig. 3.10 Temperature contours

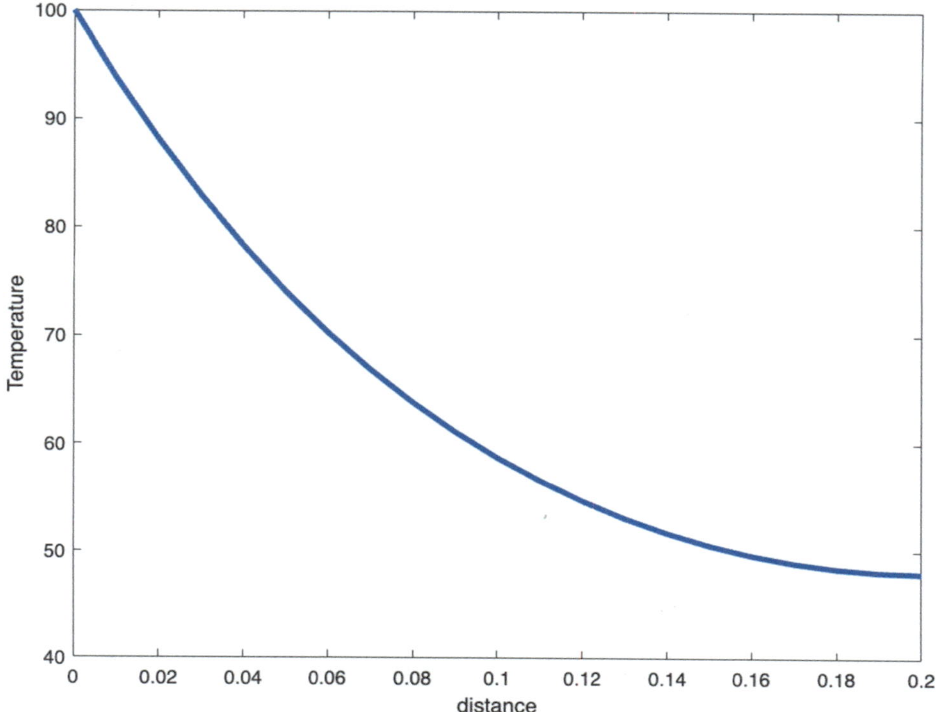

Fig. 3.11 Temperature distribution over the length

3.2 Finite Element Formulation for 2D Problems

We have followed the development of the weak form and the finite element form of typical problems in 1-D finite elements. A similar approach can be taken for 2- and 3-D problems. We will quickly summarize the process for 2-D problems here. There are many published text books on the subject which have discussed the topic in great detail.

If we consider a second order differential equation that is applicable in a 2D domain it can be written as:

$$k\left(\frac{\partial^2 T}{\partial x^2} + \frac{\partial^2 T}{\partial y^2}\right) + q = 0 \tag{3.61}$$

or

$$k\nabla^2 T + q = 0 \tag{3.62}$$

Let A be the region over which it is valid and let S be the boundary of that region.

Let S be divided into three parts S1, S2, and S3. Since this is a generic exercise we are considering all possible boundary conditions:

So the boundary conditions are specified Temperature on S1, specified flux on S2 and convective boundary condition on S3

$$T = T(x, y), on S1$$
$$\frac{\partial T}{\partial n} = f(x, y), on S2 \tag{3.63}$$
$$-k\frac{\partial T}{\partial n} = h(T - T_\infty), on S3$$

We will first derive the weak form for this equation by multiplying with a test function and integrating:

$$\overline{T}k\nabla^2 T + \overline{T}q = 0 \tag{3.64}$$

Integrating we get

$$\int_A (\overline{T}k\nabla^2 T + \overline{T}q)\overline{d}A = 0 \tag{3.65}$$

The first term in this equation can be separated into two terms using the Green-Gauss Divergence theorem,

$$\int_A k\overline{T}\nabla^2 TdA = \int_A k\overline{T}(\nabla.\nabla T)dA = -\int_A k\nabla\overline{T}\nabla TdA + \int_S k\overline{T}(\nabla T.\bar{n})ds \tag{3.66}$$

So replacing this in the original equation we get the final weak form as

$$\int_A k\nabla\overline{T}\nabla TdA = \int_A q\overline{T}dA + \int_S k\overline{T}[\nabla T.\bar{n}]dS \tag{3.67}$$

or

$$\int_A k\left[\frac{\partial T}{\partial x}\frac{\partial \overline{T}}{\partial x} + \frac{\partial T}{\partial y}\frac{\partial \overline{T}}{\partial y}\right]dA = \int_A q\overline{T}dA + \int_S k\overline{T}[\nabla T.\bar{n}]dS \tag{3.68}$$

where A is for integral over the area and S is for the integral over the surface/boundary.

In order to solve problems such as this using finite elements we need to divide the area into a large number of elements and write the integral equations over each element. Let us consider a generic linear triangular element, the basic 2-D finite element. Figure 3.12 shows such a triangle with the three vertices as the three nodes (1, 2, 3) with nodal co-ordinates at $(\times 1, y1)$, $(\times 2, y2)$ and $(\times 3, y3)$ respectively.

It can be shown that the shape functions for the triangular element is given by:

$$N_i = \frac{1}{J}(\text{pi} + \text{qix} + \text{riy}) \tag{3.69}$$

where the subscript i goes from 1 to 3, J = 2 * (area of the triangle), and
   where,

Fig. 3.12 A generic triangular element

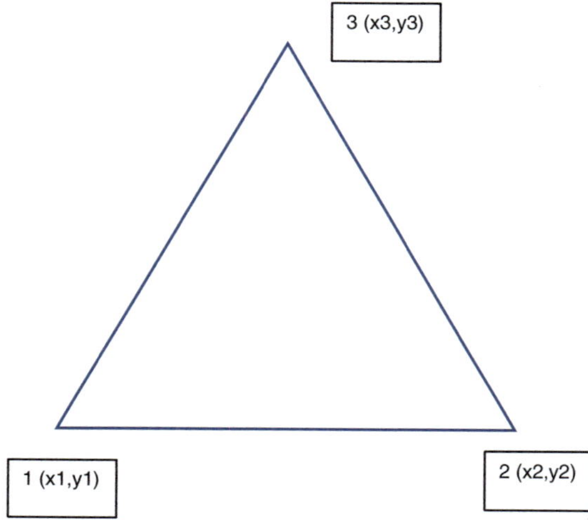

$$p1 = (x_2 y_3 - x_3 y_2); q1 = (y_2 - y_3); r1 = (x_3 - x_2) \tag{3.70}$$

$$p2 = (x_3 y_1 - x_1 y_3); q2 = (y_3 - y_1); r2 = (x_1 - x_3) \tag{3.71}$$

$$p3 = (x_1 y_2 - x_2 y_1); q3 = (y_1 - y_2); r3 = (x_2 - x_1) \tag{3.72}$$

and the primary variable,

$$u = u_1 N_1 + u_2 N_2 + u_3 N_3 \tag{3.73}$$

Example with Shape Functions

```
% if u1, u2, u3 are 100,90 and 70 respectively
% finding u somewhere inside a triangle (at (1,1))
x1 = 0;
y1 = 0;
x2 = 2;
y2 = 0;
x3 = 1;
y3 = 2;
x = 1;
y = 1;
p1 = x2*y3-x3*y2;
q1 = y2-y3;
r1 = x3-x2;
N1 = (p1 + q1*x+r1*y)/(2*2);
p2 = x1*y3-x3*y1;
q2 = y3-y1;
r2 = x1-x3;
N2 = (p2 + q2*x+r2*y)/(2*2);
p3 = x1*y2-x2*y1;
q3 = y1-y2;
r3 = x2-x1;
N3 = (p3+q3*x+r3*y)/(2*2);
u = N1*100+N2*90+N3*70
```

u = 82.5000

Using the shape function each term in the weak form of the equation results in a matrix or vector.

$$\int_A k\left[\frac{\partial T}{\partial x}\frac{\partial \overline{T}}{\partial x} + \frac{\partial T}{\partial y}\frac{\partial \overline{T}}{\partial y}\right]\mathrm{d}A = \int_A q\overline{T}\mathrm{d}A + \int_S k\overline{T}[\nabla T.\overline{n}]\mathrm{d}S \qquad (3.74)$$

The first term results in the main stiffness matrix for a triangular element where

$$K_{ij} = \frac{q_i q_j + r_i r_j}{4A} \qquad (3.75)$$

and

$$[K] = \begin{bmatrix} K_{11} & K_{12} & K_{13} \\ K_{21} & K_{22} & K_{23} \\ K_{31} & K_{32} & K_{33} \end{bmatrix} \qquad (3.76)$$

The second term in the equation results in a vector where the heat generating term q is assumed to be a constant average term on the element shown here as q^e

$$\int_A q\overline{T}\mathrm{d}A = \frac{q^e A}{3}\begin{bmatrix} 1 \\ 1 \\ 1 \end{bmatrix} \qquad (3.77)$$

The third term is an integration over the boundary. That integration on only the boundary nodes is something like this:

$$\int_S k\overline{T}(\nabla T.\overline{n})\mathrm{d}S = \frac{kl}{2}\begin{bmatrix} \frac{\partial T}{\partial n} \\ \frac{\partial T}{\partial n} \\ \frac{\partial T}{\partial n} \end{bmatrix} \qquad (3.78)$$

We will now consider an example where we apply these equations to solve a simple problem showing all the intermediate steps. Then we will consider other examples which are more complex.

Example 3.3 We will look at an example where the domain is a square domain with 0 degree temperature on the boundary. Each side is 1 unit long (Fig. 3.13). This is a symmetric problem with symmetric boundary condition so instead of considering the entire problem we can use symmetry to break it down into four symmetric regions as shown in this figure (each side is 0.5 unit for these four regions). Figure 3.14 shows the bottom left region and we can use one more level of symmetry and consider only one half of that piece using a diagonal to divide it into two as shown in Fig. 3.14.

Fig. 3.13 A unit square domain that is used in this example

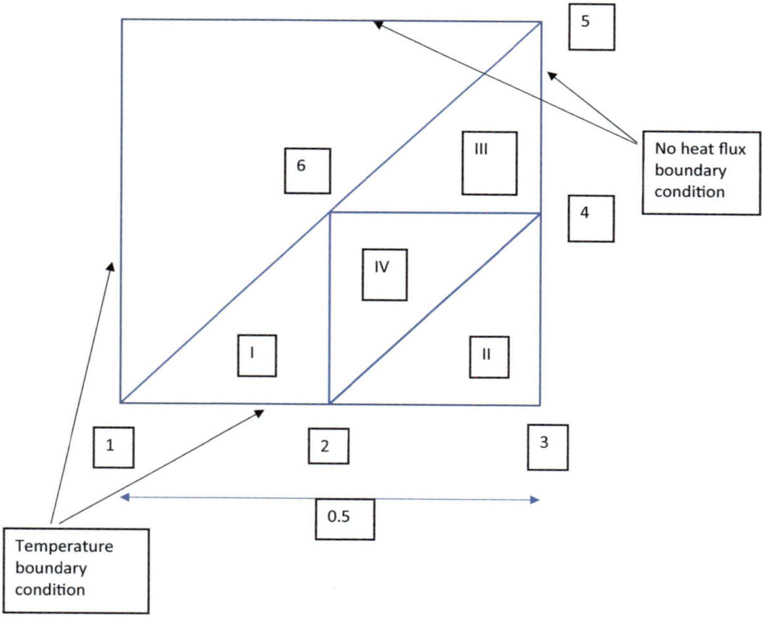

Fig. 3.14 4 finite elements to represent a slice of the domain

The Fig. 3.14 also shows how the region is divided into four triangular elements. The differential equation for heat transfer through this domain is:

$$k\left(\frac{\partial^2 T}{\partial x^2} + \frac{\partial^2 T}{\partial y^2}\right) + q = 0 \tag{3.79}$$

where T is the temperature, k is the thermal conductivity and q is the heat generation inside the domain. Let q = 1000 and k = 10.

The weak form of this function where A stands for Area and S is the outer boundary:

$$\int_A k\left[\frac{\partial T}{\partial x}\frac{\partial \overline{T}}{\partial x} + \frac{\partial T}{\partial y}\frac{\partial \overline{T}}{\partial y}\right] dA = \int_A q\overline{T}dA + \int_S k\overline{T}[\nabla T.\bar{n}]dS \tag{3.80}$$

```
Elements =
table([1;2;3;4],[1;2;6;2],[2;3;4;4],[6;4;5;6],'VariableNames',{'NEL','Node1','No
de2','Node3'})
```

Elements = 4×4 table

	NEL	Node1	Node2	Node3
1	1	1	2	6
2	2	2	3	4
3	3	6	4	5
4	4	2	4	6

```
Nodes =
table([1;2;3;4;5;6],[0;0.25;0.5;0.5;0.5;0.25],[0;0;0;0.25;0.5;0.25],'VariableNam
es',{'Node','X-Coordinate','Y-Coordinate'})
```

Nodes = 6×3 table

	Node	X-Coordinate	Y-Coordinate
1	1	0	0
2	2	0.2500	0
3	3	0.5000	0
4	4	0.5000	0.2500
5	5	0.5000	0.5000
6	6	0.2500	0.2500

Using this information, we create the stiffness matrix from the following relationships:

$$N_i = \frac{1}{J}(\text{pi} + \text{qix} + \text{riy}) \tag{3.81}$$

were,

$$\text{p1} = (x_2 y_3 - x_3 y_2); \text{q1} = (y_2 - y_3); \text{r1} = (x_3 - x_2) \tag{3.82}$$

$$p2 = (x_3 y_1 - x_1 y_3); q2 = (y_3 - y_1); r2 = (x_1 - x_3) \tag{3.83}$$

$$p3 = (x_1 y_2 - x_2 y_1); q3 = (y_1 - y_2); r3 = (x_2 - x_1) \tag{3.84}$$

We saw earlier that for the stiffness matrix the elements are given by:

$$K_{ij} = \frac{q_i q_j + r_i r_j}{4A} \tag{3.85}$$

For Element 1 (Fig. 3.15)

$$K_{11} = \frac{\left(-\frac{1}{4}\right)^2 + 0}{4\left(\frac{1}{2}\right)\left(\frac{1}{4}\right)^2} = \frac{1}{2}; K_{12} = \frac{\left(-\frac{1}{4}\right)\left(\frac{1}{4}\right) + 0}{4\left(\frac{1}{2}\right)\left(\frac{1}{4}\right)^2} = -\frac{1}{2}; K_{12} = \frac{\left(-\frac{1}{4}\right)(0) + 0}{4\left(\frac{1}{2}\right)\left(\frac{1}{4}\right),^2} = 0 \tag{3.86}$$

$$K_{22} = \frac{\left(-\frac{1}{4}\right)^2 + \left(-\frac{1}{2}\right)^2}{4\left(\frac{1}{2}\right)\left(\frac{1}{4}\right)^2} = 1; K_{23} = \frac{(0)\left(\frac{1}{4}\right) + \left(-\frac{1}{4}\right)\left(\frac{1}{4}\right)}{4\left(\frac{1}{2}\right)\left(\frac{1}{4}\right)^2} = -\frac{1}{2}; K_{22} = \frac{1}{2} \tag{3.87}$$

For Elements 2 and 3 the matrices are the same.

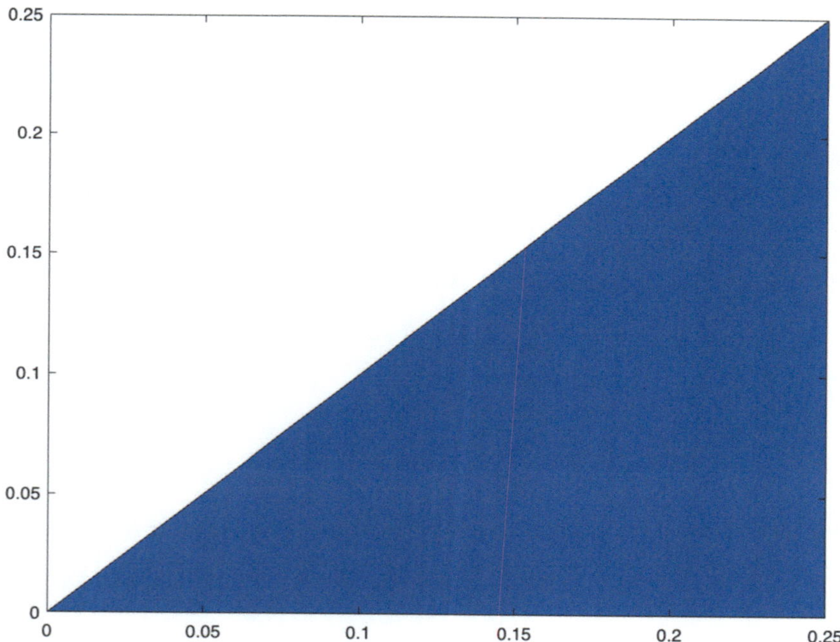

Fig. 3.15 Element 1 from this example

$$[K]^1 = \frac{k}{2} \begin{bmatrix} 1 & -1 & 0 \\ -1 & 2 & -1 \\ 0 & -1 & 1 \end{bmatrix} \begin{bmatrix} T_1 \\ T_2 \\ T_6 \end{bmatrix}; \ [K]^2 = \frac{k}{2} \begin{bmatrix} 1 & -1 & 0 \\ -1 & 2 & -1 \\ 0 & -1 & 1 \end{bmatrix} \begin{bmatrix} T_2 \\ T_3 \\ T_4 \end{bmatrix};$$

$$[K]^3 = \frac{k}{2} \begin{bmatrix} 1 & -1 & 0 \\ -1 & 2 & -1 \\ 0 & -1 & 1 \end{bmatrix} \begin{bmatrix} T_6 \\ T_4 \\ T_5 \end{bmatrix}$$

(3.88)

For element 4 the matrix is slightly different as shown below

$$[K]^4 = \frac{k}{2} \begin{bmatrix} 1 & 0 & -1 \\ 0 & 1 & -1 \\ -1 & -1 & 2 \end{bmatrix} \begin{bmatrix} T_2 \\ T_4 \\ T_6 \end{bmatrix}$$

(3.89)

The right-hand side term becomes for each triangle

$$\int_A q\overline{T}\mathrm{dA} = \frac{q^e A}{3} \begin{bmatrix} 1 \\ 1 \\ 1 \end{bmatrix}$$

(3.90)

$$\int_S k\overline{T}(\nabla T.\overline{n})\mathrm{dS} = \frac{\mathrm{kl}}{2} \begin{bmatrix} \frac{\partial T}{\partial n} \\ \frac{\partial T}{\partial n} \\ \frac{\partial T}{\partial n} \end{bmatrix};$$

(3.91)

This is a general vector but it only applies to nodes on the boundary. We will show this in the final assembled matrix.

When we assemble the matrices.

$$\frac{k}{2} \begin{bmatrix} 1 & -1 & 0 & 0 & 0 & 0 \\ -1 & (2+1+1) & -1 & 0 & 0 & (-1-1) \\ 0 & -1 & 2 & -1 & 0 & 0 \\ 0 & 0 & -1 & (1+2+1) & -1 & (0-1) \\ 0 & 0 & 0 & -1 & 1 & -1 \\ 0 & (-1-1) & 0 & -1 & -1 & (2+1+1) \end{bmatrix} \begin{bmatrix} T_1 \\ T_2 \\ T_3 \\ T_4 \\ T_5 \\ T_6 \end{bmatrix}$$

(3.92)

$$= \frac{(1000)\left(\frac{1}{2}\left(\frac{1}{4}\frac{1}{4}\right)\right)}{3} \begin{bmatrix} 1 \\ 1+1+1 \\ 1 \\ 1+1+1 \\ 1 \\ 1+1+1 \end{bmatrix} + \frac{\left(10.\frac{1}{4}\right)}{2} \begin{bmatrix} 0 \\ 0 \\ \frac{\partial T}{\partial n} \\ \frac{\partial T}{\partial n} \\ \frac{\partial T}{\partial n} \\ 0 \end{bmatrix}$$

Since temperatures at nodes 1, 2, and 6 are unknowns and temperatures at odes 3, 4 and 5 are equal to 0. We can separate the first second and sixth equation and solve for the unknowns.

$$\frac{10}{2}\begin{bmatrix} 1 & -1 & 0 \\ -1 & 4 & -2 \\ 0 & -2 & 4 \end{bmatrix}\begin{bmatrix} T_1 \\ T_2 \\ T_6 \end{bmatrix} = 10.4167\begin{bmatrix} 1 \\ 3 \\ 3 \end{bmatrix} \tag{3.93}$$

```
A = [1 -1 0;-1 4 -2;0 -2 4]

A = 3×3
     1    -1     0
    -1     4    -2
     0    -2     4
```

```
b = 10.4167/5*[1;3;3]

b = 3×1
    2.0833
    6.2500
    6.2500
```

```
T = A\b

T = 3×1
    7.8125
    5.7292
    4.4271
```

So T1 = 7.8125, T2 = 5.7292, T3 = 4.4271.

And the three derivatives at the outer boundary are calculated in the following fashion.

```
K = [1 -1 0 0 0 0;-1 4 -1 0 0 -2;0 -1 2 -1 0 0;0 0 -1 4 -1 -1;0 0 0 -1 1 -1;0 -2 0 -1 -1 4]

K = 6×6
     1    -1     0     0     0     0
    -1     4    -1     0     0    -2
     0    -1     2    -1     0     0
     0     0    -1     4    -1    -1
     0     0     0    -1     1    -1
     0    -2     0    -1    -1     4
```

```
T = [7.8125;5.7292;0;0;0;4.4271]

T = 6×1
    7.8125
    5.7292
         0
         0
         0
    4.4271
```

```
b = 10.4167/5*[1;3;1;3;1;3]

b = 6×1
     2.0833
     6.2500
     2.0833
     6.2500
     2.0833
     6.2500

(K*T-b)/1.125

ans = 6×1
    -0.0000
     0.0001
    -6.9445
    -9.4908
    -5.7871
    -0.0000

x = [0 0.25];
y = [0 0.25];
area(x,y)
hold on
```

So, on each triangle the derivative can be calculated by taking the derivative of the temperature approximation:

$$\frac{\partial T}{\partial x} = T1\frac{\partial N1}{\partial x} + T2\frac{\partial N2}{\partial x} + T3\frac{\partial N3}{\partial x} = (T1(q1) + T2(q2) + T3(q3))\frac{1}{J} \qquad (3.94)$$

and

$$\frac{\partial T}{\partial y} = T1\frac{\partial N1}{\partial y} + T2\frac{\partial N2}{\partial y} + T3\frac{\partial N3}{\partial y} = (T1(r1) + T2(r2) + T3(r3))\frac{1}{J} \qquad (3.95)$$

where,

$$N_i = \frac{1}{J}(pi + qix + riy) \qquad (3.96)$$

and,

$$p1 = (x_2y_3 - x_3y_2); q1 = (y_2 - y_3); r1 = (x_3 - x_2) \qquad (3.97)$$

$$p2 = (x_3y_1 - x_1y_3); q1 = (y_3 - y_1); r1 = (x_1 - x_3) \qquad (3.98)$$

$$p3 = (x_1y_2 - x_2y_1); q1 = (y_1 - y_2); r1 = (x_2 - x_1) \qquad (3.99)$$

So, for element 1 the x and y derivatives of T in the element are:
The x derivative of the temperature is written as:

$$\frac{\partial T}{\partial x} = (T1(q1) + T2(q2) + T3(q3))\frac{1}{J}$$

$$= \frac{1}{2\left(\frac{1}{2}\right)\left(\frac{1}{4}\right)^2}\{7.8125(0-0.25) + 5.7292(0.25-0) + 4.4271(0)\} \tag{3.100}$$

$$= 16\{-0.5208\} = -8.332$$

$$\frac{\partial T}{\partial y} = (T1(r1) + T2(r2) + T3(r3))\frac{1}{J} = \frac{1}{2\left(\frac{1}{2}\right)\left(\frac{1}{4}\right)^2}\{7.8125(0) + 5.7292(0-0.25) + 4.4271(0.25-0)\}$$
$$= 16\{-0.3255\} = -5.2084$$

$$\tag{3.101}$$

We will look at several examples to explore some of the features of typical problems in continuum heat transfer such as, domains that have different material properties, time dependent or transient problems, as well as internal and external heat sources.

Example 3.4 This problem is a steady state heat transfer problem where the wall of a refrigerator is made of fiberglass insulation (k = 0.035 W/m K) sandwiched between two layers of 1 mm thick sheet metal (k = 15.1 W/m K). The refrigerated space is maintained at a temperature of 3 C. The average heat transfer coefficients of the inner and outer surface of the wall are 4 W/m^2 K and 9 W/m^2 K respectively. The kitchen temperature averages 25 C. It is observed that if outer surface of the refrigerator drops below 20 C condensation forms. Find the minimum thickness of fiberglass needed to avoid this (Fig. 3.16).

Thermal Analysis: Compute Temperature Distribution

Create a steady state thermal model. Assume the thickness of the insulation to be 5 mm for this first trial.

```
modelforfridge = createpde('thermal','steadystate'); %model
```

Create a geometry with two adjacent rectangles. The top edge of the longer rectangle (on the right) represents the disc-pad contact region.

```
R1 = [3,4, [ 0, 1,  1,    0, -10, -10, 0, 0]/1000]'; %definition of the first
region(inner layer of sheet metal)
R2 = [3,4, [1, 6, 6, 1, -10, -10, 0, 0]/1000]';%defination of the second region
(fiber glass insulation)
R3 = [3,4, [6, 7, 7, 6, -10,-10, 0, 0]/1000]';%definition of the third region
(outer layer of sheet metal)

gdm = [R1 R2 R3];
ns = char('R1','R2','R3');
g = decsg(gdm,'R1 + R2 + R3',ns'); %geometry definition as a combination of the
three regions
```

Assign the geometry to the thermal model.

```
geometryFromEdges(modelforfridge,g);
```

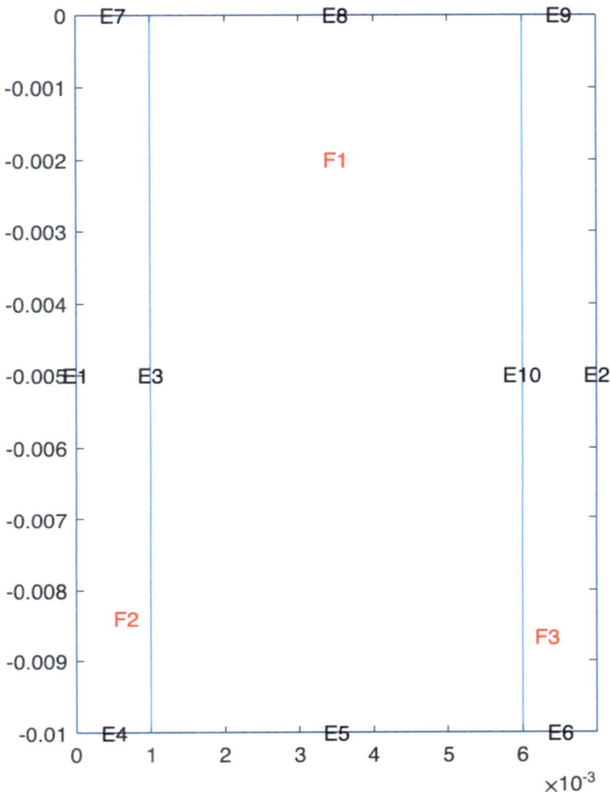

Fig. 3.16 Composite geometry for Example 3.4

Plot the geometry with the edge and face labels (Fig. 3.16)

```
figure
pdegplot(modelforfridge,'EdgeLabels','on','FaceLabels','on')
```

Generate a mesh. For the mesh we use the linear geometric order (Fig. 3.17).

```
generateMesh(modelforfridge,'Hmax',0.3E-03,'GeometricOrder','linear');
figure
pdeplot(modelforfridge,"Mesh","on")
axis equal
```

Specify the thermal material properties of the three parts of the domain.

```
Kd1 = 15.1; %conductivity of region 1
Kd2 = 0.035; %conductivity of region 2
Kd3 = 15.1; %conductivity of region 3
thermalProperties(modelforfridge,'face', 2, 'ThermalConductivity',Kd1);
thermalProperties(modelforfridge,'face', 1, 'ThermalConductivity',Kd2);
thermalProperties(modelforfridge,'face', 3, 'ThermalConductivity',Kd3);
```

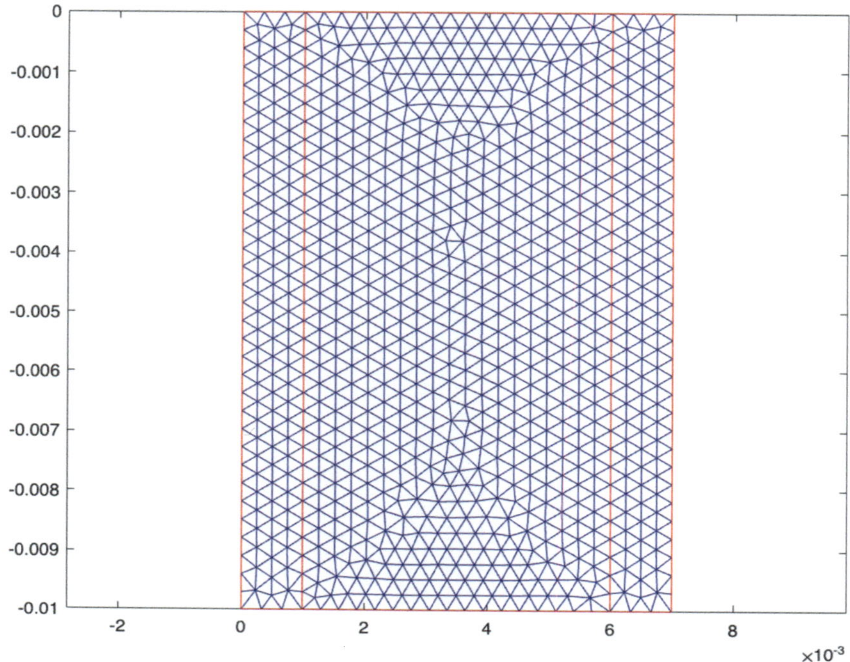

Fig. 3.17 Meshed geometry for Example 3.4

Specify the convective boundary condition at the two ends.

```
thermalBC(modelforfridge,'Edge',1,'ConvectionCoefficient',4,'AmbientTemperature
',3); % the condition at the left end
thermalBC(modelforfridge,'Edge',2,'ConvectionCoefficient',9,'AmbientTemperature
',25);% the condition at the right end

Rt = solve(modelforfridge);% solve the problem and assign the results to the
variable Rt
T = Rt.Temperature; % Temperature
flux = Rt.evaluateHeatFlux; % Heat Flux
figure
pdeplot(modelforfridge,'XYData',T,'Contour','on','ColorBar','on','ColorMap','je
t')
axis equal
title 'Steady-State Temperature'
```

```
figure
pdeplot(modelforfridge,'XYData',flux,'Contour','on','ColorBar','on','ColorMap',
'jet')
axis equal
title 'Heat Flux'
```

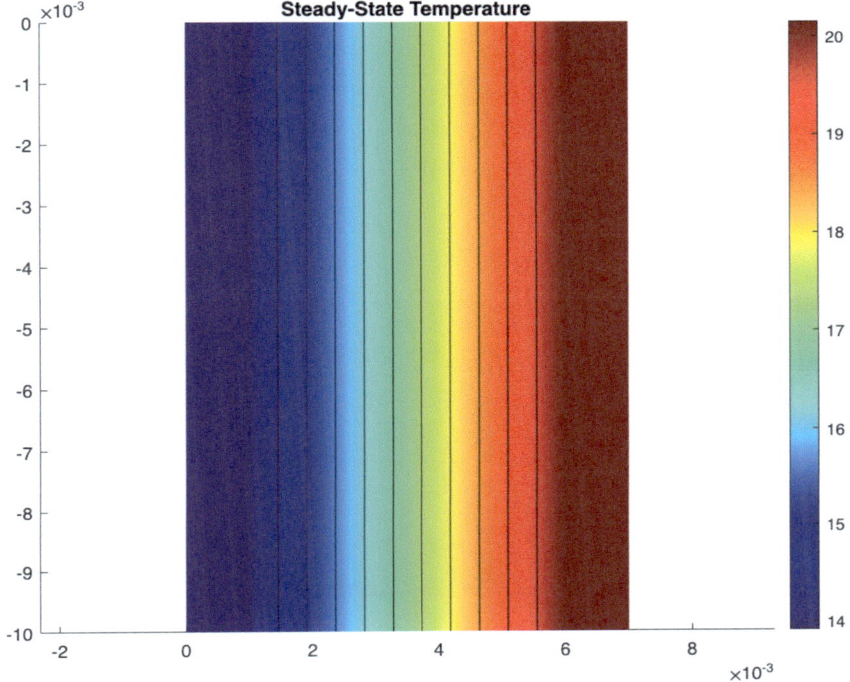

Fig. 3.18 Temperature contours

Plot the temperature variation across the length of the domain (Fig. 3.18) and heat flux (Fig. 3.19).

```
%plotting the temperature across the length of the bar
X = 0:0.001:0.011;
Y = zeros(size(X));

Tintrp = interpolateTemperature(Rt,X,Y);

plot(X,Tintrp,"LineWidth",3);
xlabel('Distance');
ylabel('Temperature')
```

The results show that the temperature at the outer wall is just above 20 C (Fig. 3.20). So the thickness of 5 mm is just about sufficient for this application. We could re-do this with a smaller thickness and check if that thickness will be sufficient to meet the design requirements or not.

Example 3.5 We will now consider a heat transfer problem that is transient. The temperature profile will vary with time. There are many real-life problems where heat transfer is a function of time. We will consider a representative problem where the geometry has a L-shape and there is a hole inside a portion of the geometry (Fig. 3.21). One edge of the

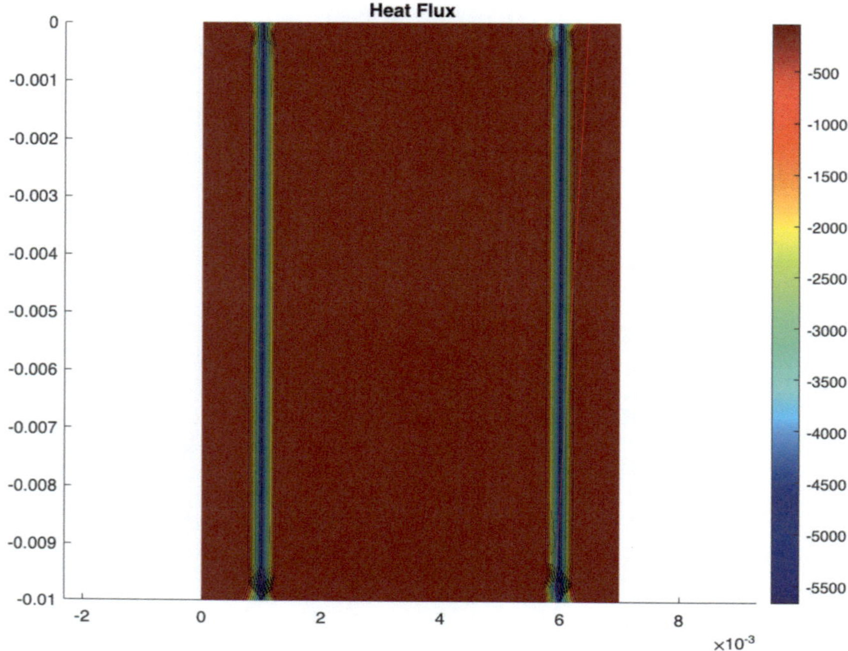

Fig. 3.19 Heat Flux contours

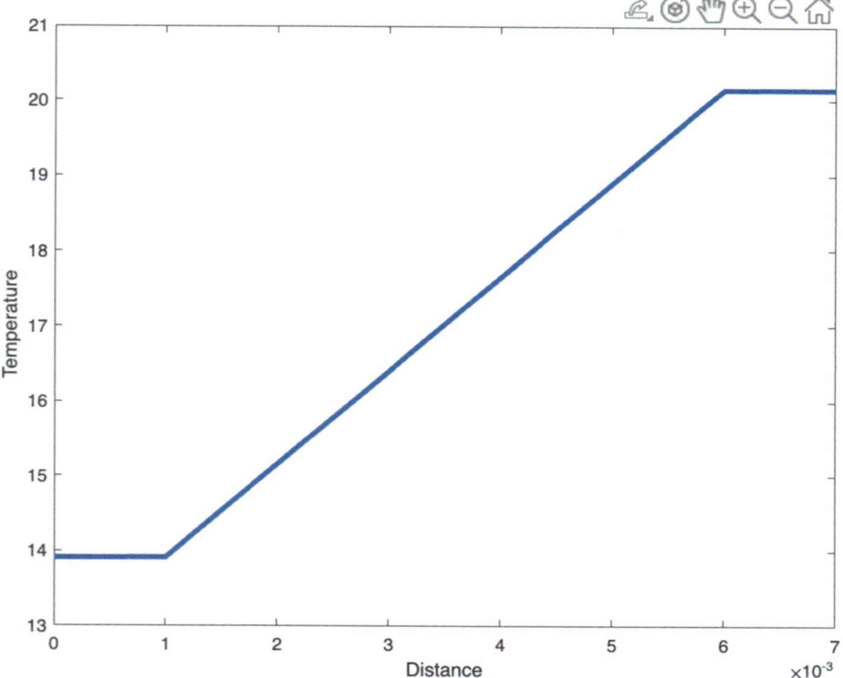

Fig. 3.20 Temperature variation across the length

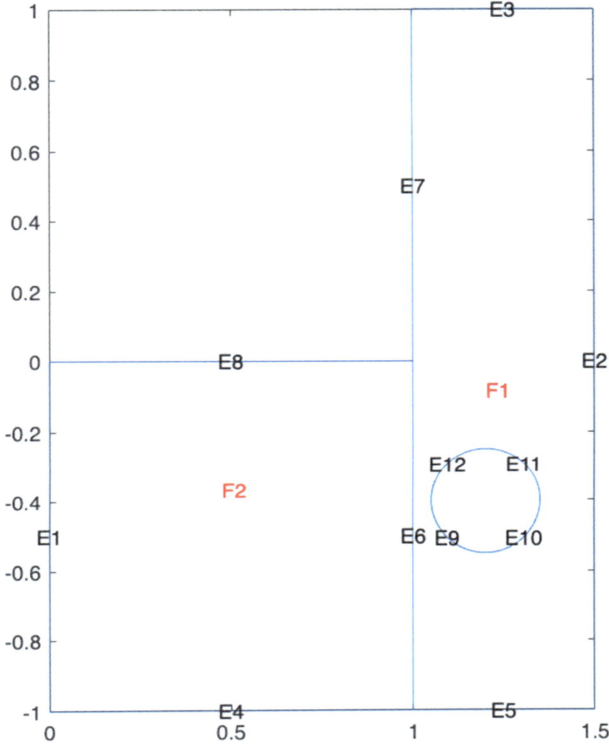

Fig. 3.21 Geometry of Example 3.5

geometry is held at 100 C. The other edges have convective boundary conditions with convection coefficient as 10 W/m^2 K and the ambient temperature is 25 C. The material properties are conductivity = 51 W/m K, density = 7100 kg/m^3, specific heat = 50 J/kg K.

```
modelforL = createpde('thermal','transient'); %model and soluttion type is
defined
R1 = [3,4, [ 0, 10,  10,   0, -10, -10, 0, 0]/10]'; % rectangle 1
R2 = [3,4, [10, 15, 15, 10, -10, -10, 10, 10]/10]'; % rectangle 2
C1 = [1, [12, -4, 1.5,0,0,0,0,0,0]/10]'; %circle

gdm = [R1 R2 C1];
ns = char('R1','R2','C1');
g = decsg(gdm,'R1 + R2 - C1',ns');% total geeometry is R1+R2-C1
geometryFromEdges(modelforL,g); % geometry is made using basic shapes
```

Plot the geometry with the edge and face labels.

```
figure
pdegplot(modelforL,'EdgeLabels','on','FaceLabels','on') %composite geometry is
drawn
```

Generate a mesh, using linear geometric order instead of the default quadratic order (Fig. 3.22).

```
generateMesh(modelforL,'Hmax',0.6E-01,'GeometricOrder','linear'); % the mesh
size is defined and liner elements are used
figure
pdeplot(modelforL,"Mesh","on") % plot the mesh
axis equal
```

Specify the thermal material properties of the domain.

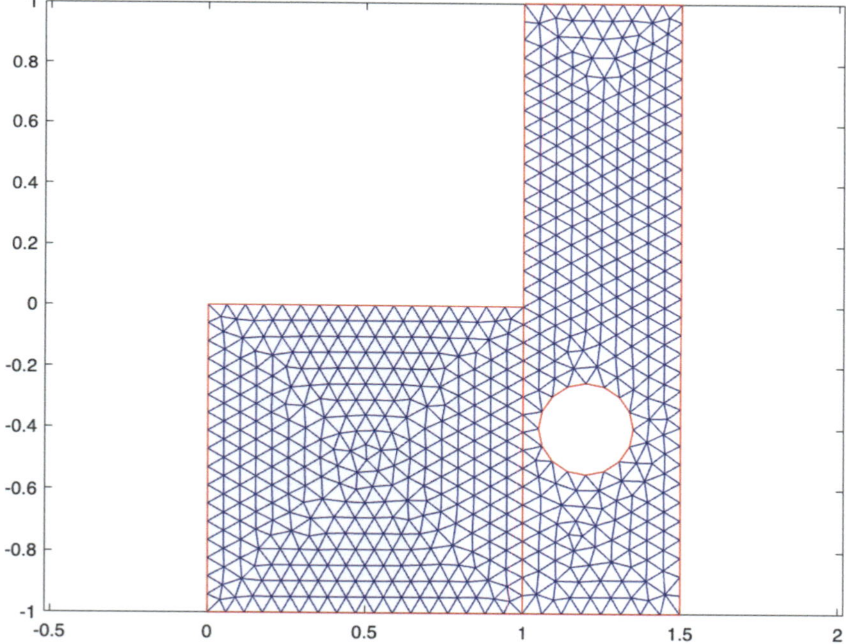

Fig. 3.22 Meshed geometry for Example 3.5

```matlab
Kd = 51; %conductivity
rho = 7100;% mass density
cp = 50; % specific heat
thermalProperties(modelforL,
'ThermalConductivity',Kd,'MassDensity',rho,'SpecificHeat',cp); % assigning
properties

thermalIC(modelforL,25); %assigning initial conditions for the entire domain
%Thermal Boundary Conditions
thermalBC (modelforL,'Edge',1,'Temperature',100);
thermalBC(modelforL,'Edge',4,'ConvectionCoefficient',10,'AmbientTemperature',25
);
 thermalBC(modelforL,'Edge',5,'ConvectionCoefficient',10,'AmbientTemperature',25
);
 thermalBC(modelforL,'Edge',2,'ConvectionCoefficient',10,'AmbientTemperature',25
);
 thermalBC(modelforL,'Edge',3,'ConvectionCoefficient',10,'AmbientTemperature',25
);
 thermalBC(modelforL,'Edge',7,'ConvectionCoefficient',10,'AmbientTemperature',25
);
 thermalBC(modelforL,'Edge',9,'ConvectionCoefficient',10,'AmbientTemperature',25
);
 thermalBC(modelforL,'Edge',10,'ConvectionCoefficient',10,'AmbientTemperature',2
5);
 thermalBC(modelforL,'Edge',11,'ConvectionCoefficient',10,'AmbientTemperature',2
5);
 thermalBC(modelforL,'Edge',12,'ConvectionCoefficient',10,'AmbientTemperature',2
5);
```

Specify Solution Times.

Set solution times to be 0–10,000 s in steps of 1.

```matlab
tlist = 0:1:10000;% the sollution is attempted for 10000 seconds
```

Calculate Solution.

Use the `solve` function to calculate the solution.

```matlab
Rt = solve(modelforL,tlist); % solve the problem

pdeplot(modelforL,'XYData',Rt.Temperature(:,1000), ...
                  'Contour','on','ColorMap','jet') % plot the temperature
contours at 1000 seconds
 axis equal
```

Temperature contours at 1000 sec and at 10,000 sec are shown in Figs. 3.23 and 3.24, respectively. Figure 3.25 shows the temperature vs. time profile at three different locations in the domain.

```matlab
pdeplot(modelforL,'XYData',Rt.Temperature(:,end), ...
                  'Contour','on','ColorMap','jet')%plot the temperature
contours after 10000 seconds
 axis equal
```

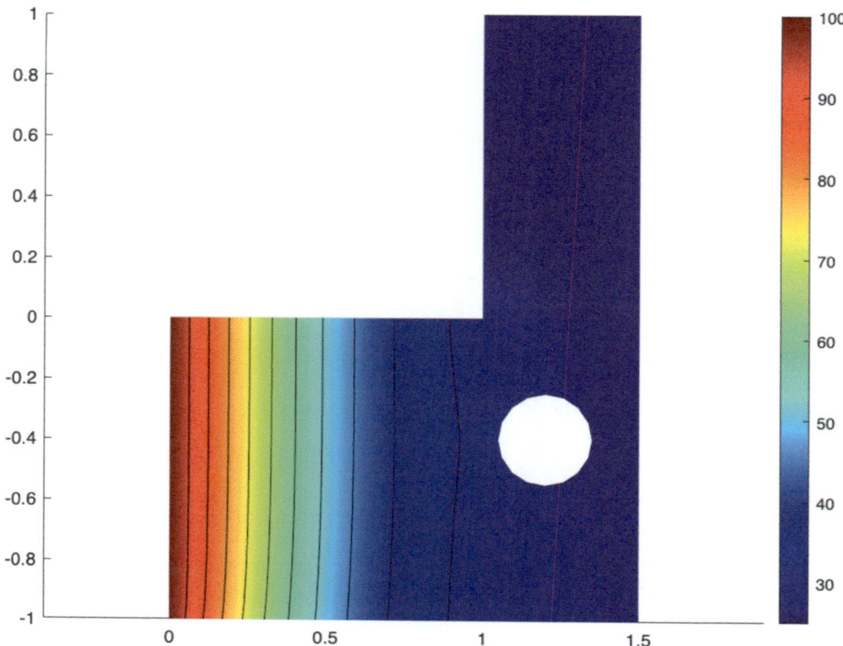

Fig. 3.23 Temperature contours after 1000 s

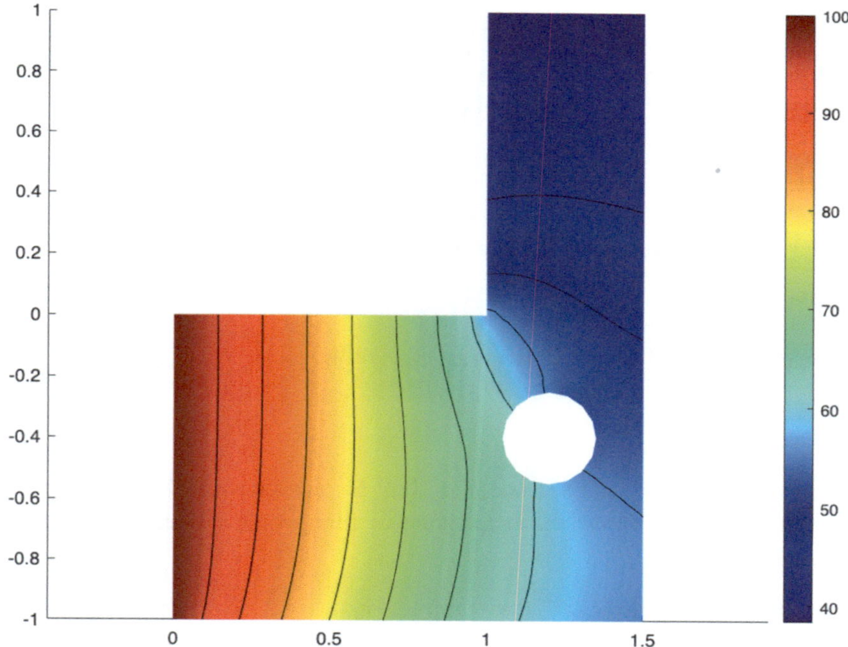

Fig. 3.24 Temperature contours after 10,000 s

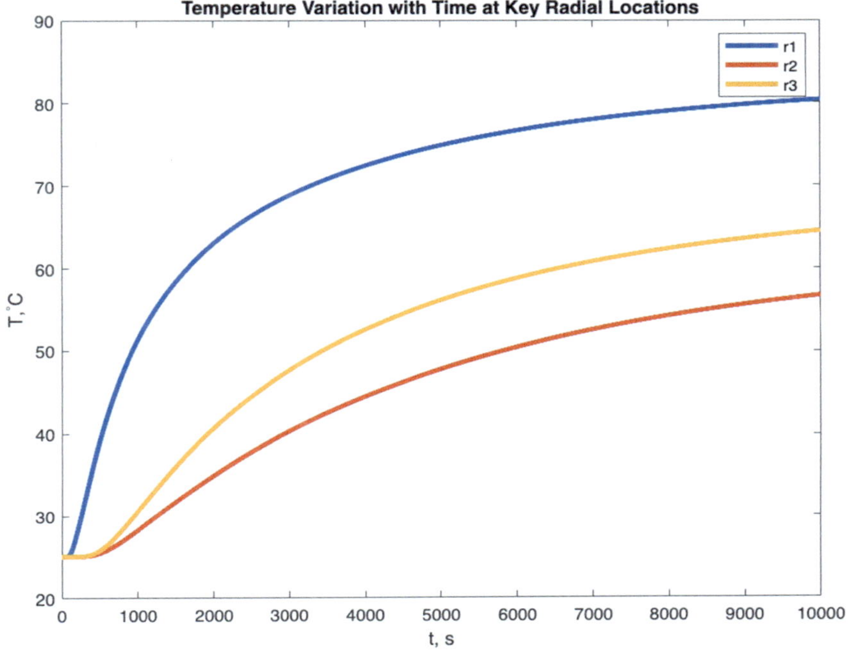

Fig. 3.25 Temperature versus time at three locations in the domain

```
% plot the time Vs. temperature at three selected locations.
% At r1(0.5,-0.4), r2(1,0) and r3(1,-0.4)
iTr1 = interpolateTemperature(Rt,[0.5;-0.4],1:numel(Rt.SolutionTimes));
iTr2 = interpolateTemperature(Rt,[1;0],1:numel(Rt.SolutionTimes));
iTr3= interpolateTemperature(Rt,[1;-0.4],1:numel(Rt.SolutionTimes));


figure
plot(tlist,iTr1,'LineWidth',3)
hold on
plot(tlist,iTr2,'LineWidth',3)
plot(tlist,iTr3,'LineWidth',3)
title('Temperature Variation with Time at Key Radial Locations')
legend('r1','r2','r3')
xlabel 't, s'
ylabel 'T,^{\circ}C'
```

Example 3.5 (with quadratic elements)

We now run the same exact example with quadratic elements but make them
approximately double the size.

```
modelforL = createpde('thermal','transient'); %model and soluttion type is
defined
R1 = [3,4, [ 0, 10,  10,   0, -10, -10, 0, 0]/10]'; % rectangle 1
R2 = [3,4, [10, 15, 15, 10, -10, -10, 10, 10]/10]'; % rectangle 2
C1 = [1, [12, -4, 1.5,0,0,0,0,0,0]/10]'; %circle

gdm = [R1 R2 C1];
ns = char('R1','R2','C1');
g = decsg(gdm,'R1 + R2 - C1',ns');% total geeometry is R1+R2-C1
geometryFromEdges(modelforL,g); % geometry is made using basic shapes
```

Plot the geometry with the edge and face labels (Fig. 3.26).

```
figure
pdegplot(modelforL,'EdgeLabels','on','FaceLabels','on') %composite geometry is
drawn
```

Generate a mesh, using linear geometric order instead of the default quadratic order
(Fig. 3.27).

```
generateMesh(modelforL,'Hmax',1.2E-01,'GeometricOrder','quadratic'); % the mesh
size is defined and liner elements are used
figure
pdeplot(modelforL,"Mesh","on") % plot the mesh
axis equal
```

Specify the thermal material properties of the domain.

```
Kd = 51; %conductivity
rho = 7100;% mass density
cp = 50; % specific heat
thermalProperties(modelforL,
'ThermalConductivity',Kd,'MassDensity',rho,'SpecificHeat',cp); % assigning
properties

thermalIC(modelforL,25); %assigning initial conditions for the entire domain
%Thermal Boundary Conditions
thermalBC (modelforL,'Edge',1,'Temperature',100);
thermalBC(modelforL,'Edge',4,'ConvectionCoefficient',10,'AmbientTemperature',25
);
thermalBC(modelforL,'Edge',5,'ConvectionCoefficient',10,'AmbientTemperature',25
);
thermalBC(modelforL,'Edge',2,'ConvectionCoefficient',10,'AmbientTemperature',25
);
thermalBC(modelforL,'Edge',3,'ConvectionCoefficient',10,'AmbientTemperature',25
);
thermalBC(modelforL,'Edge',7,'ConvectionCoefficient',10,'AmbientTemperature',25
);
thermalBC(modelforL,'Edge',9,'ConvectionCoefficient',10,'AmbientTemperature',25
);
thermalBC(modelforL,'Edge',10,'ConvectionCoefficient',10,'AmbientTemperature',2
5);
thermalBC(modelforL,'Edge',11,'ConvectionCoefficient',10,'AmbientTemperature',2
5);
thermalBC(modelforL,'Edge',12,'ConvectionCoefficient',10,'AmbientTemperature',2
5);
```

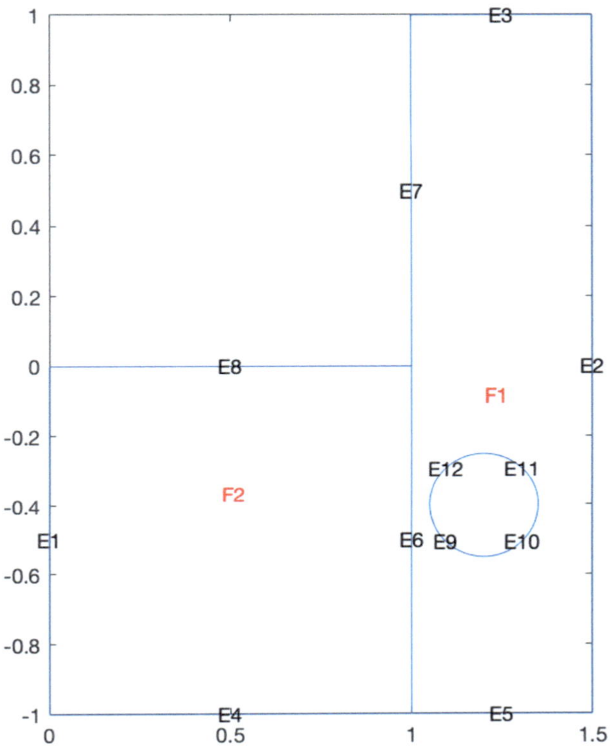

Fig. 3.26 Geometry of Example 3.5

Specify Solution Times.
Set solution times to be 0–10,000 s in steps of 1.

```
tlist = 0:1:10000;% the sollution is attempted for 10000 seconds
```

Calculate Solution.
Use the `solve` function to calculate the solution.

```
Rt = solve(modelforL,tlist); % solve the problem

pdeplot(modelforL,'XYData',Rt.Temperature(:,1000), ...
                'Contour','on','ColorMap','jet') % plot the temperature
contours at 1000 seconds
 axis equal

pdeplot(modelforL,'XYData',Rt.Temperature(:,end), ...
                'Contour','on','ColorMap','jet')%plot the temperature
contours after 10000 seconds
 axis equal
```

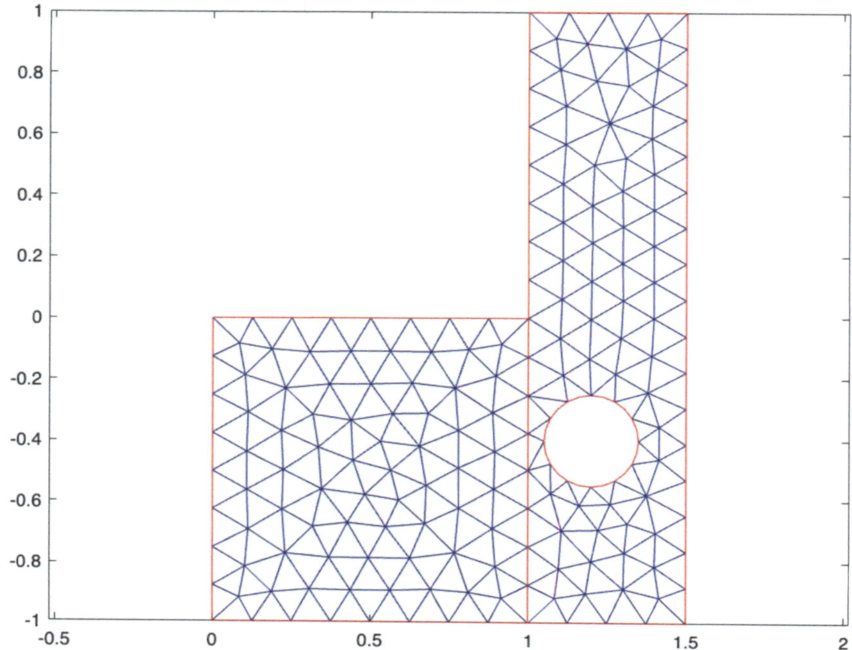

Fig. 3.27 Quadratic mesh with fewer elements

```
% plot the time Vs. temperature at three selected locations.
% At r1(0.5,-0.4), r2(1,0) and r3(1,-0.4)
iTr1 = interpolateTemperature(Rt,[0.5;-0.4],1:numel(Rt.SolutionTimes));
iTr2 = interpolateTemperature(Rt,[1;0],1:numel(Rt.SolutionTimes));
iTr3= interpolateTemperature(Rt,[1;-0.4],1:numel(Rt.SolutionTimes));


figure
plot(tlist,iTr1,'LineWidth',3)
hold on
plot(tlist,iTr2,'LineWidth',3)
plot(tlist,iTr3,'LineWidth',3)
title('Temperature Variation with Time at Key Radial Locations')
legend('r1','r2','r3')
xlabel 't, s'
ylabel 'T,^{\circ}C'
```

Figures 3.28 and 3.29 show the temperature profile calculated using quadratic elements at 1000 sec and 10000 sec, respectively. One can compare these with Figs. 3.23 and 3.24 respectively. Also, Fig. 3.30 shows the temperature variation with time at the three locations that were plotted in Fig. 3.26 from the linear analysis. In Example 3.5 we saw

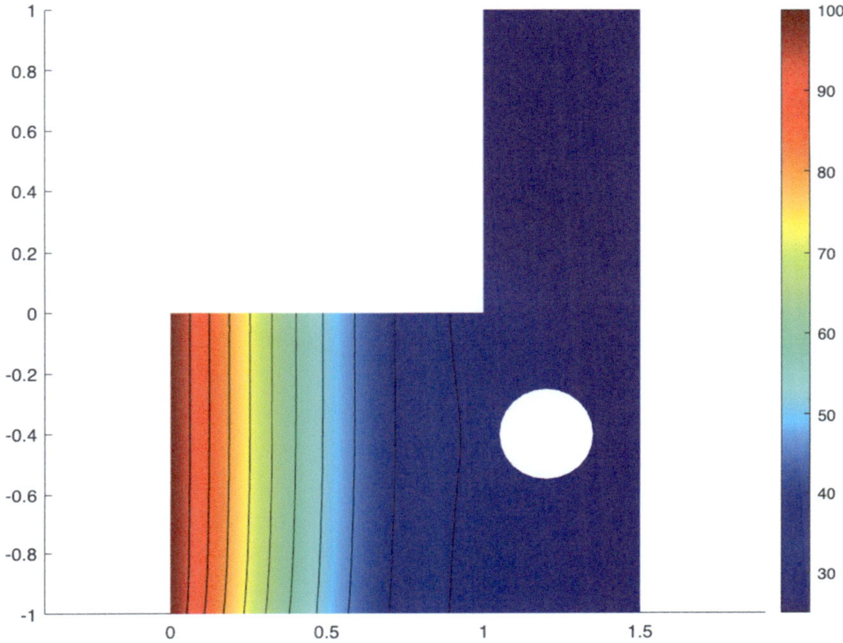

Fig. 3.28 Temperature contours after 1000 s

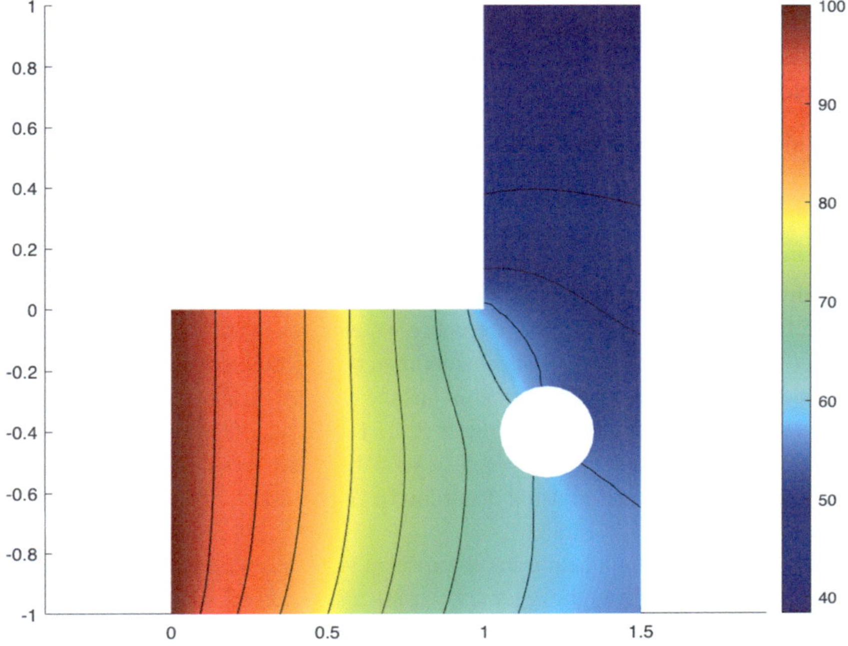

Fig. 3.29 Temperature contours after 10,000 s

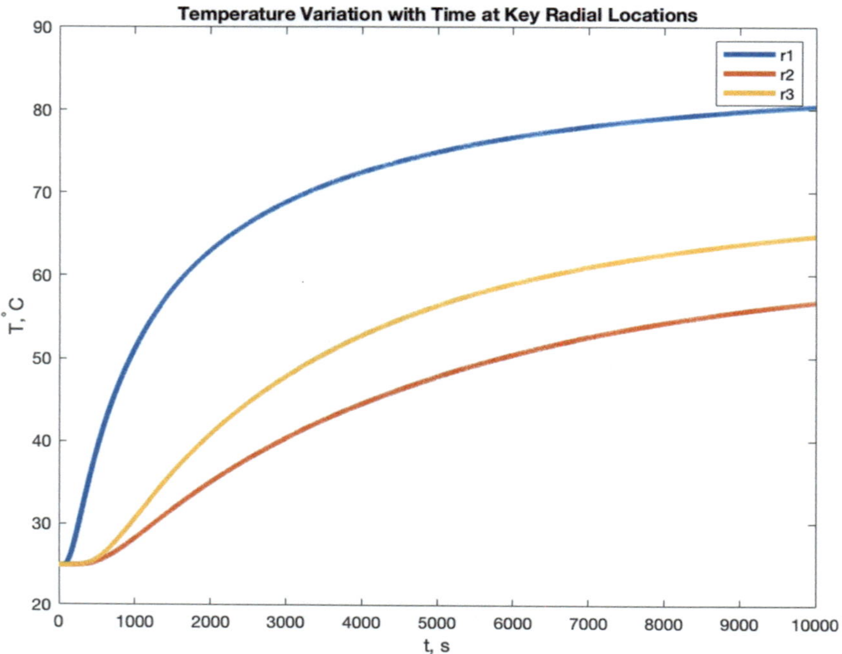

Fig. 3.30 Temperature versus time at three locations in the domain

that analyzing the problem first with linear elements (and more elements) and then with quadratic elements (with fewer elements) yields essentially the same results as seen in the contour plots and plots of temperature versus time.

Example 3.6 Watermelons are often cut in half and put in freezers to cool quickly. But usually we forget to check on it and end up having a watermelon with a frozen layer on the top. To avoid this potential problem a person wants to set the timer such that it will go off when the temperature of the exposed surface of the watermelon drops to 3 °C.

Consider a watermelon with two different diameters. The watermelon is cut in half and put in the freezer at −12 C. Initially, the entire watermelon is at a uniform temperature of 25 C. And the convective heat transfer coefficient from the surface is 22 W/m^2 C. We assume in this analysis that the watermelon has the properties of water and analyze for the time it would take for the center to reach 3 C. Density = 1000 kg/m^3, specific heat = 4187 J/kg K, thermal conductivity = 0.598 W/m K. In the model all the dimensions are in mm so the properties are adjusted accordingly (Fig. 3.31).

Create Thermal Analysis Model.

The first step in solving a heat transfer problem is to create a thermal analysis model. This time we are importing the geometry from a .stl file of the 3D geometry of the watermelon. The model created will contain the geometry, thermal material properties,

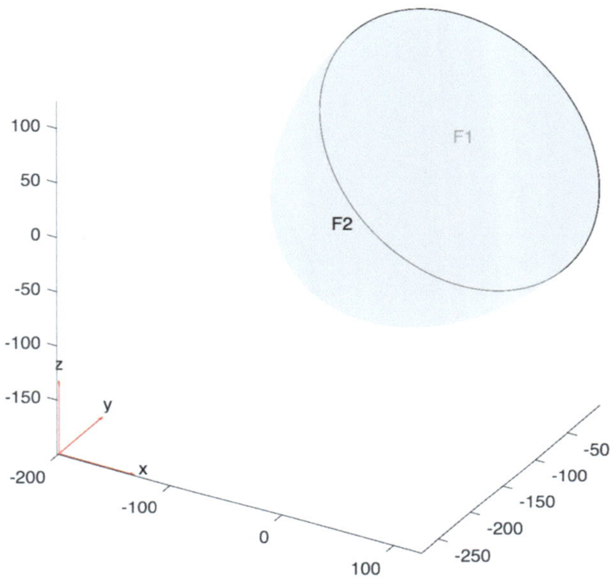

Fig. 3.31 Watermelon geometry

internal heat sources, temperature on the boundaries, heat fluxes through the boundaries, mesh, and initial conditions.

```
therwater = createpde('thermal','transient'); % a transient heat transfer model
is used
importGeometry(therwater,"watermelon v2.stl");% geometry in mm is imported via
a stl file
figure
pdegplot(therwater,"FaceLabels","on","EdgeLabels","on","FaceAlpha",0.5)%draw
the geometry with faces and edges shown
```

Specify Thermal Properties of Material.
Specify the thermal conductivity, mass density, and specific heat of the material.

```
thermalProperties(therwater,'ThermalConductivity',0.000598,...
                            'MassDensity',1E-6,...
                            'SpecificHeat',4187);%necessary material
properties
```

Apply Boundary Conditions.
Specify the temperature on the left edge as `100`, and constant heat flow to the exterior through the right edge as `-10`. The toolbox uses the default insulating boundary condition for all other boundaries.

```
thermalBC(therwater,'Face',[1 2],'ConvectionCoefficient',22E-
6,'AmbientTemperature',-12);
% applying convective Boundary conditions
```

Fig. 3.32 Meshed geometry

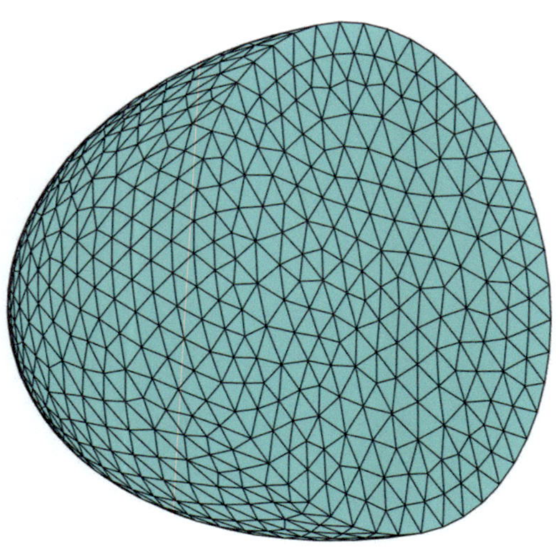

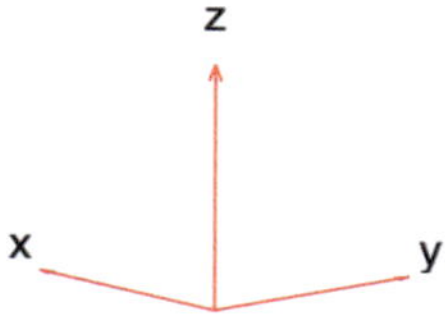

Set Initial Conditions.

Set an initial value of 0 for the temperature.

```
thermalIC(therwater,25); %applying inital conditions
```

Generate Mesh (Fig. 3.32).

Create and plot a mesh.

```
generateMesh(therwater,"GeometricOrder","linear");
figure
pdemesh(therwater)
title('Mesh with Linear Triangular Elements')

view([132 -12])
```

Specify Solution Times.

Set solution times to be 0–6000 s in steps of 1/2 s.

```
tlist = 0:0.5:6000;
```

Calculate Solution.

Use the `solve` function to calculate the solution.

```
thermalresults = solve(therwater,tlist); %store the results in "thermalresults"
```

Plot Temperature Distribution and Heat Flux.

Plot the solution contour at three different times, t = 50, 500 and 5000 s (Figs. 3.33, 3.34 and 3.35).

```
pdeplot3D(therwater,"ColorMapData",thermalresults.Temperature(:,101))
colorbar('ticks',[-5 0 5 10 15 20 25],'TickLabels',{'-
5','0','5','10','15','20','25'})

view([112 9])
title ('Temperature distribution after 50 sec')


pdeplot3D(therwater,"ColorMapData",thermalresults.Temperature(:,1001))
colorbar('ticks',[-5 0 5 10 15 20 25],'TickLabels',{'-
5','0','5','10','15','20','25'})

view([112 9])
title ('Temperature distribution after 500 sec')
```

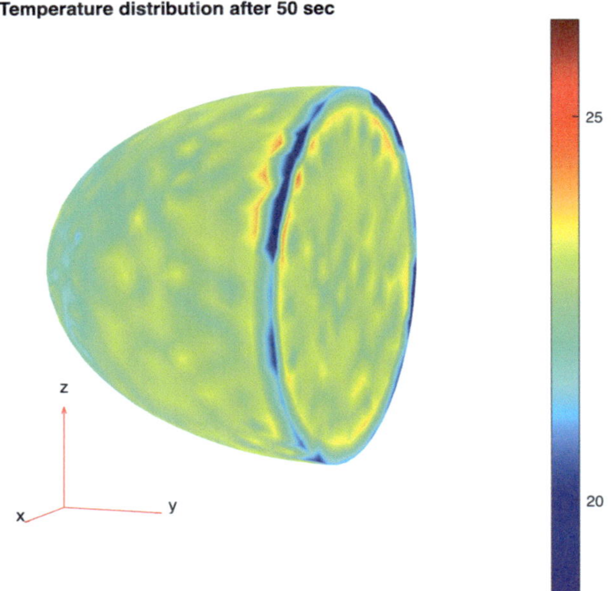

Fig. 3.33 Temperature distribution at 50 s

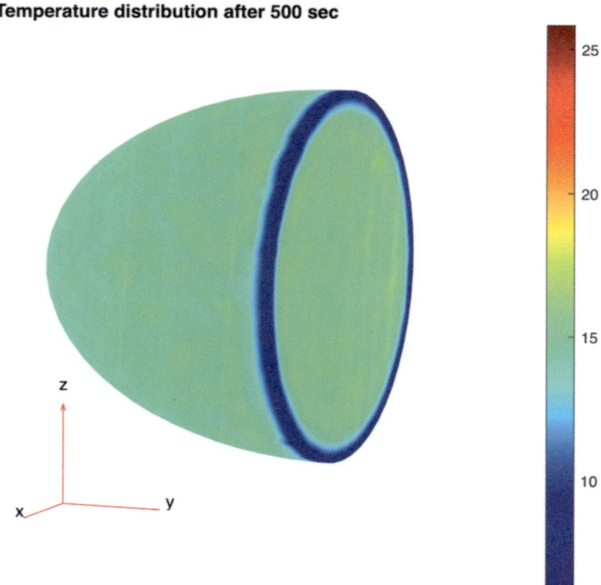

Fig. 3.34 Temperature distribution at 500 s

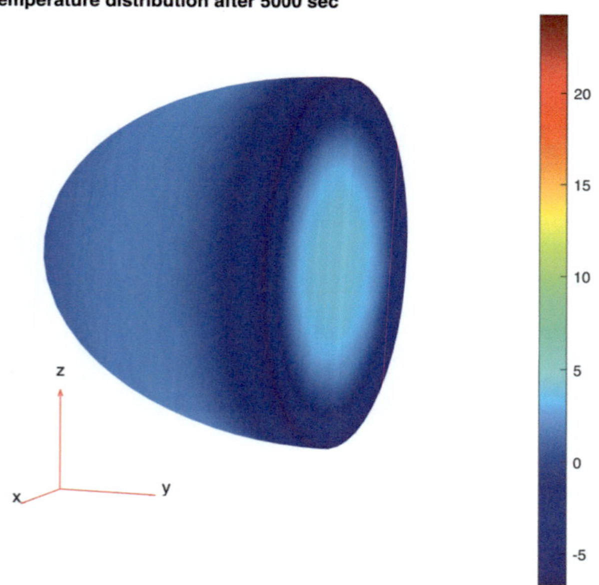

Fig. 3.35 Temperature distribution at 5000 s

```matlab
pdeplot3D(therwater,"ColorMapData",thermalresults.Temperature(:,10001))
colorbar('ticks',[-5 0 5 10 15 20 25],'TickLabels',{'-
5','0','5','10','15','20','25'})

view([112 9])
title ('Temperature distribution after 5000 sec')

pdemesh(therwater,"NodeLabels","on","FaceAlpha",0.8)% to show the node numbers
on the face of the watermelon
view([132 -12])

xlim([-107 31])
ylim([-197 -79])
zlim([-106 31])
view([155 9])

view([157 -5])
```

Plotting the Temperature versus time results for three locations around the center of the watermelon to determine the time around when the temperature reaches 3 °C (Fig. 3.36).

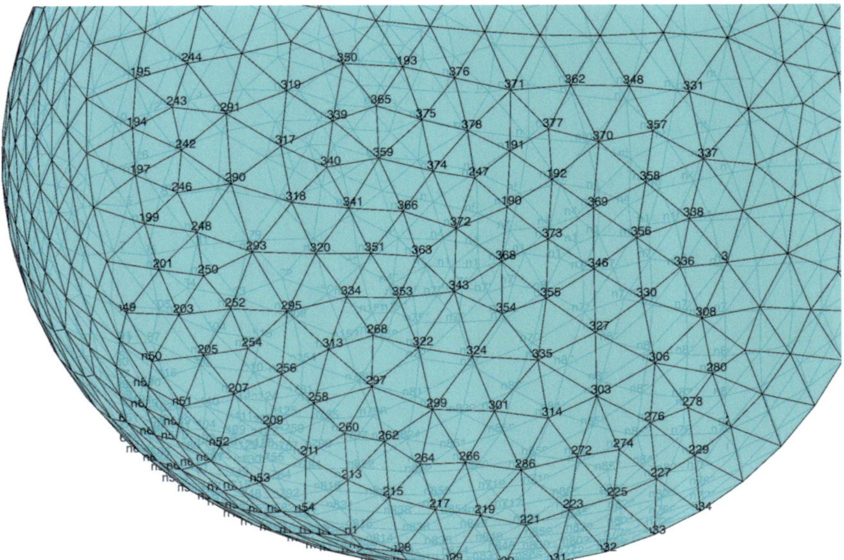

Fig. 3.36 Closeup of meshed geometry to show node numbers

```
figure
plot(thermalresults.SolutionTimes,thermalresults.Temperature(378,:))
hold on
plot(thermalresults.SolutionTimes,thermalresults.Temperature(371,:))
plot(thermalresults.SolutionTimes,thermalresults.Temperature(191,:))
title('Temperature Variation with Time at Center of the section')
legend('P_1','P_2','P_3')
xlabel 't, s'
ylabel 'T,^{\circ}C'
grid on
```

It is clear from this plot (Fig. 3.37) that the temperature at the center reaches 3 °C at around 6000 s which is around an hour and a half.

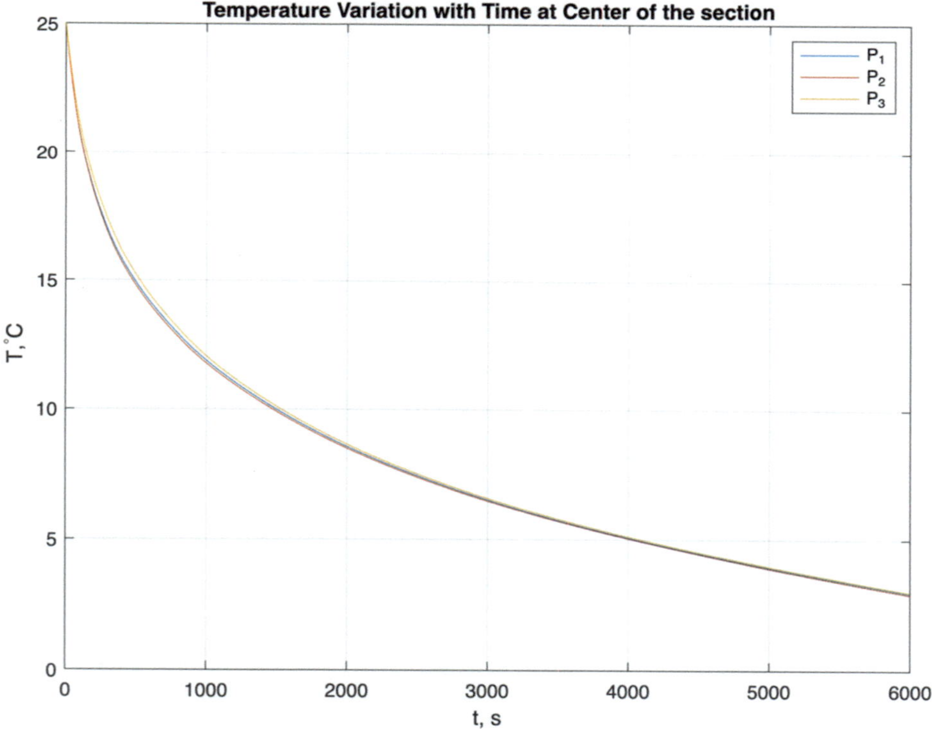

Fig. 3.37 Temperature change over time for three points around the center of the watermelon

3.3   Summary

In this chapter we explored the basics of continuum FEA first through 1-D problems and then through some 2-D and 3-D problems. We also looked at conditions such as layered material with different material properties, steady state and transient problems, use of linear and quadratic elements as well as importing geometry (using .stl files) created with other CAD tools. All the discussion was done using heat transfer examples since the primary values in heat transfer problems is a scalar or uni-dimensional at a point. In other words, at any point there is only one variable, the temperature. In the next chapter we will explore stress analysis problems where the primary variable at any point will have multiple components.

Stress Analysis Problems **4**

In this chapter we explore a number of problem types that have to do with stress analysis and mechanically induced deformation. We will consider both static and dynamic problems. We will first look at Beams, the FEA formulation for beam problems and how that is implemented for different loading and boundary conditions, followed by a more automated or efficient model to solve beam problems. After that we consider the formulation for plane elasticity problems and a more automated version of a static problem. Then we consider a problem to track the dynamic behavior of a structure leading to oscillatory motion.

4.1 Beams

Beams are structural elements that are used in numerous structural applications and the characteristic that separates beams from other structural elements is that the beams support transverse forces/loads. Typically beams have one dimension (length) that is significantly larger than the other dimensions (cross-section) and the load that is supported by beams act transverse or perpendicular to its length.

Typical load types that are supported by a beam are: (a) Concentrated forces (b) Concentrated moments, (c) various types of distributed forces.

The differential equation for beam bending is a 4th order equation:

$$\frac{d^2}{dx^2}\left(\text{EI}\,\frac{d^2u}{dx^2}\right) = f(x) \tag{4.1}$$

where u is the beam displacement, f(x) is the distributed load and EI is the product of Young's modulus and the 2nd moment of area of the beam crosss-section.

S. Das, *An Introduction to Finite Element Analysis Using Matlab Tools*,
Synthesis Lectures on Mechanical Engineering,
https://doi.org/10.1007/978-3-031-17540-4_4

$$u = \text{\textit{displacement}}$$

$$\frac{du}{dx} = \theta = \text{\textit{slope}}$$

$$\left(EI\frac{d^2u}{dx^2} \right) = M(x) = \text{\textit{Bending Moment}} \tag{4.2}$$

$$-\frac{d}{dx}\left(EI\frac{d^2u}{dx^2} \right) = V(x) = \text{\textit{Shear Force}}$$

The sign conventions commonly used for Shear force and bending moment are as follows:

Shear force is positive when pointing up and negative when pointing down. Bending moment is positive when it is clockwise (looking at the right side of location of the bending moment) and is negative when it is counter clockwise (looking at the right side of the bending moment).

To set up the above differential equation for a finite element formulation we multiply the equation with a test function w(x) and integrate the two sides. Particularly the term involving the derivatives is integrated twice by parts:

$$\int w\frac{d^2}{dx^2}\left(EI\frac{d^2u}{dx^2} \right)dx = \int wfdx \tag{4.3}$$

$$\left[w\frac{d}{dx}\left(EI\frac{d^2u}{dx^2} \right) \right] - \int \frac{dw}{dx}\frac{d}{dx}\left(EI\frac{d^2u}{dx^2} \right)dx = \int wfdx \tag{4.4}$$

$$\left[w\frac{d}{dx}\left(EI\frac{d^2u}{dx^2} \right) \right] - \left[w\left(EI\frac{d^2u}{dx^2} \right) \right] + EI\int \frac{d^2w}{dx^2}\left(EI\frac{d^2u}{dx^2} \right)dx = \int wfdx \tag{4.5}$$

In order to solve this weak form using finite elements we divide each beam into multiple number of elements. So for any generic 1-D element if the end points of the element are at $\times 1$ and $\times 2$, we can write the element equation as:

$$\left[w\frac{d}{dx}\left(EI\frac{d^2u}{dx^2} \right) \right]_{x_1}^{x_2} - \left[w\left(EI\frac{d^2u}{dx^2} \right) \right]_{x_1}^{x_2} + EI\int_{x_1}^{x_2} \frac{d^2w}{dx^2}\left(EI\frac{d^2u}{dx^2} \right)dx \neq \int_{x_1}^{x_2} wfdx \tag{4.6}$$

The first term involves shear force and the second term involves bending moment. The inequality is used instead of an equal sign because these terms in Eq. 4.6 are written for a single element. The equality will only hold when the all the elements are assembled into a set of coupled linear equations.

We can re-write this as:

$$EI\int_{x_1}^{x_2} \frac{d^2w}{dx^2}\left(EI\frac{d^2u}{dx^2} \right)dx \neq \int_{x_1}^{x_2} wfdx - \left[w\frac{d}{dx}\left(EI\frac{d^2u}{dx^2} \right) \right]_{x_1}^{x_2} + \left[w\left(EI\frac{d^2u}{dx^2} \right) \right]_{x_1}^{x_2} \tag{4.7}$$

$$\text{EI} \int_{x_1}^{x_2} \frac{d^2w}{dx^2} \left(\text{EI} \frac{d^2u}{dx^2} \right) dx \neq \int_{x_1}^{x_2} \text{wfdx} + [\text{wV}(x)]_{x_1}^{x_2} + [\text{wM}(x)]_{x_1}^{x_2} \tag{4.8}$$

since,

$$\left(EI \frac{d^2u}{dx^2} \right) = M(x) = Bending\ Moment$$

$$-\frac{d}{dx} \left(EI \frac{d^2u}{dx^2} \right) = V(x) = Shear\ Force \tag{4.9}$$

Beam Elements

Beam elements need to be somewhat different from spring elements even though both are 1-D elements. For the spring elements the displacement at the end points need to be continus. In the case of the beam both the displacement and the slope need to be continuous at the end points for each elements, i.e. these two quantities need to be continuous between two adjacent elements.

So, the general equation for the beam element is written as:

$$u = u_1 N_1 + \theta_1 N_2 + u_2 N_3 + \theta_2 N_4 \tag{4.10}$$

$$N_1 = 1 - \frac{3x^2}{L^2} + \frac{2x^3}{L^3} \tag{4.11}$$

$$N_2 = x - \frac{2x^2}{L} + \frac{x^3}{L^2} \tag{4.12}$$

$$N_3 = \frac{3x^2}{L^2} - \frac{2x^3}{L^3} \tag{4.13}$$

$$N_4 = -\frac{x^2}{L} + \frac{x^3}{L^2} \tag{4.14}$$

Applying this to the integral term on the left hand side of the above equation we can generate the standard stiffness matrix of a single beam element:

$$\frac{EI}{L^3} \begin{bmatrix} 12 & 6L & -12 & 6L \\ 6L & 4L^2 & -6L & 2L^2 \\ -12 & -6L & 12 & -6L \\ 6L & 2L^2 & -6L & 4L^2 \end{bmatrix} \tag{4.15}$$

Using this the element equation can be written as:

$$\frac{EI}{L^3}\begin{bmatrix} 12 & 6L & -12 & 6L \\ 6L & 4L^2 & -6L & 2L^2 \\ -12 & -6L & 12 & -6L \\ 6L & 2L^2 & -6L & 4L^2 \end{bmatrix}\begin{bmatrix} u_1 \\ \theta_1 \\ u_2 \\ \theta_2 \end{bmatrix} \neq \begin{bmatrix} V_1 \\ M_1 \\ V_2 \\ M_2 \end{bmatrix} + [\,distributed\,force\,effect\,] \quad (4.16)$$

The inequality sign is used since without assembly the left hand side and the right hand side are not quite equal to each other.

It can be shown that the distributed force term can become equal to:

$$\frac{f_e L}{12}\begin{bmatrix} 6 \\ L \\ 6 \\ -L \end{bmatrix} \quad (4.17)$$

f_e is an equivalent distributed load (a single average value) for the element. We will elaborate more on this in examples. When the elements are assembled the sum of shear forces from adjacent elements is equal to the external Shear force (or concentrated force) and the sum of bending moments from two adjacent elements is equal to the externally applied moment.

$$V_2^1 + V_1^2 = V \quad (4.18)$$

$$M_2^1 + M_1^2 = M \quad (4.19)$$

Example 4.1 Consider a cantilever beam with a uniformly distributed load. Its length is L and the uniformly distributed load is w and since it is applied downwards we will use $-$ w. For this exercise we will use just one element to represent the beam. We will use EI as a constant (Fig. 4.1).

For the single element representing the beam:

$$\frac{EI}{L^3}\begin{bmatrix} 12 & 6L & -12 & 6L \\ 6L & 4L^2 & -6L & 2L^2 \\ -12 & -6L & 12 & -6L \\ 6L & 2L^2 & -6L & 4L^2 \end{bmatrix}\begin{bmatrix} u_1 \\ \theta_1 \\ u_2 \\ \theta_2 \end{bmatrix} = \begin{bmatrix} V_1 \\ M_1 \\ V_2 \\ M_2 \end{bmatrix} + \frac{f_e L}{12}\begin{bmatrix} 6 \\ L \\ 6 \\ -L \end{bmatrix} \quad (4.20)$$

u_1 and θ_1 are zero since the left end is fixed.

Fig. 4.1 A cantilever beam with distributed load

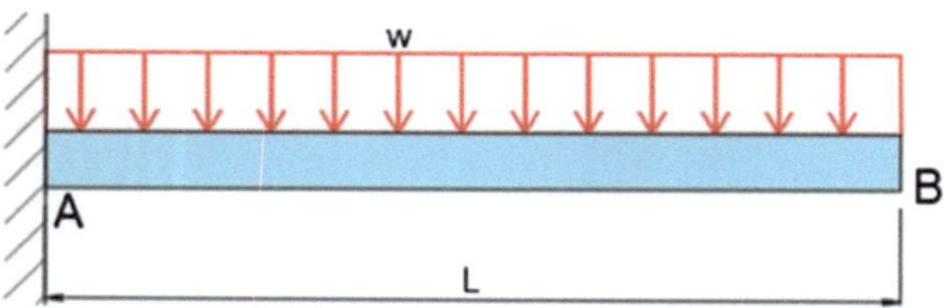

So if we apply that condition the top two equations can drop off. Also V_2 and M_2 are zero, because there is no shear force or bending moment at the right end. So, if we take the last two equations we can solve for u_2 and θ_2. And then if we substitute that result in the original four equations we can solve for V_1 and M_1:

$$\frac{EI}{L^3}\begin{bmatrix} 12 & -6L \\ -6L & 4L^2 \end{bmatrix}\begin{bmatrix} u_2 \\ \theta_2 \end{bmatrix} = \frac{-wL}{12}\begin{bmatrix} 6 \\ -L \end{bmatrix} = \begin{bmatrix} \frac{-wL}{2} \\ \frac{wL^2}{12} \end{bmatrix} \tag{4.21}$$

$$\begin{bmatrix} 12 & -6L \\ -6L & 4L^2 \end{bmatrix}\begin{bmatrix} u_2 \\ \theta_2 \end{bmatrix} = \begin{bmatrix} \frac{-wL^4}{2EI} \\ \frac{wL^5}{12EI} \end{bmatrix} \tag{4.22}$$

```
% Calculatinng deflections
L = sym('L');
w = sym('w');
EI = sym('EI');
A = [12 -6*L;-6*L 4*L^2];
f = [-w*L^4/(2*EI); w*L^5/(12*EI)];
soln = A\f;
u2 = soln(1)

u2 =
 L^4 w
-────
 8 EI

theta2 = soln(2)

theta2 =
 L^3 w
-────
 6 EI

% Calculating Reactions at the wall
A = EI/L^3*[12 6*L -12 6*L;6*L 4*L^2 -6*L 2*L^2;-12 -6*L 12 -6*L;
    6*L 2*L^2 -6*L 4*L^2];
f = -w*L/12*[6;L;6;-L];
u = [0;0;u2;theta2]

u =

   0
   0
  L^4 w
 -────
  8 EI
  L^3 w
 -────
  6 EI

V = A*u-f

V =

  L w
  L^2 w
  ────
   2
   0
   0
```

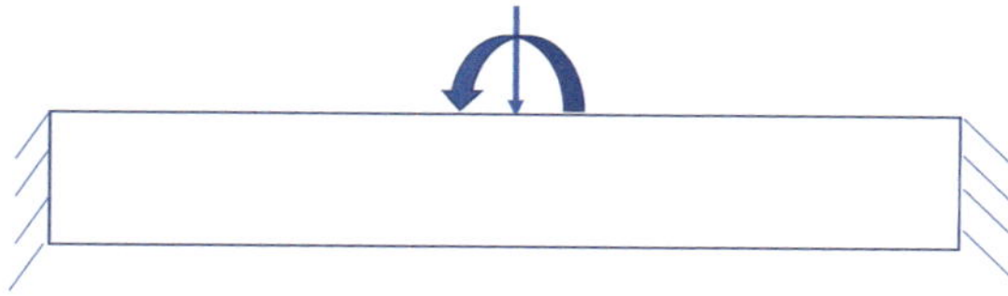

Fig. 4.2 Loaded beam with fixed ends

Example 4.2 This is a beam that is fixed at both ends. It is 6 m long. At the midpoint there is a downward concentrated force of 10 kN and a counterclockwise moment of 20 kNm. There is no distributed load. The beam is divided into two elements of 3 m length. E = 210E9 Pa, I = 4E−4 m^4 (Fig. 4.2).

```
clear
L = 3;
w = 0;
EI = 210E9*4E-4;
% Definnition of Stiffness Matrices
kel = EI/L^3*[12 6*L -12 6*L;6*L 4*L^2 -6*L 2*L^2;-12 -6*L 12 -6*L;
    6*L 2*L^2 -6*L 4*L^2]
```

kel = 4×4
10^8 ×

```
      0.3733      0.5600     -0.3733      0.5600
      0.5600      1.1200     -0.5600      0.5600
     -0.3733     -0.5600      0.3733     -0.5600
      0.5600      0.5600     -0.5600      1.1200
```

```
fel = w*L/12*[6;L;6;-L];
V2 = -10000
```

V2 = -10000

```
M2 = 20000
```

```
M2 = 20000
```

```
% Initializing
elem_con = [1 2; 2 3]; %elemment connectivity
elem = size(elem_con,1);%total nummber of elements
nodes = elem+1;% in a 1 D system #nodes = #elements+1
K = zeros(2*nodes);%Initialize global stiffness matrix
f = zeros(2*nodes,1)% initialize the distributed load term
```

```
f = 6×1
      0
      0
      0
      0
      0
      0
```

```
%Assembly
for i=1:elem
    %ke = k(i)*[1 -1;-1 1]
    node11 = elem_con(i,1)+(i-1)*1
    node12 = elem_con(i,1)+(i-1)*1+1
    node21 = elem_con(i,2)+(i-1)*1+1
    node22 = elem_con(i,2)+(i-1)*1+2
    K(node11,node11) = K(node11,node11) + kel(1,1);
    K(node11,node12) = K(node11,node12) + kel(1,2);
    K(node11,node21) = K(node11,node21) + kel(1,3);
    K(node11,node22) = K(node11,node22) + kel(1,4);
    %
    K(node12,node11) = K(node12,node11) + kel(2,1);
    K(node12,node12) = K(node12,node12) + kel(2,2);
    K(node12,node21) = K(node12,node21) + kel(2,3);
    K(node12,node22) = K(node12,node22) + kel(2,4);
    %
    K(node21,node11) = K(node21,node11) + kel(3,1);
    K(node21,node12) = K(node21,node12) + kel(3,2);
    K(node21,node21) = K(node21,node21) + kel(3,3);
    K(node21,node22) = K(node21,node22) + kel(3,4);
    %
    K(node22,node11) = K(node22,node11) + kel(4,1);
    K(node22,node12) = K(node22,node12) + kel(4,2);
    K(node22,node21) = K(node22,node21) + kel(4,3);
    K(node22,node22) = K(node22,node22) + kel(4,4);
end
```

```
node11 = 1
node12 = 2
node21 = 3
node22 = 4
node11 = 3
node12 = 4
node21 = 5
node22 = 6
```

```
% Create the Distributed load vector
for i=1:elem
    node11 = elem_con(i,1)+(i-1)*1;
    node12 = elem_con(i,1)+(i-1)*1+1;
    node21 = elem_con(i,2)+(i-1)*1+1;
    node22 = elem_con(i,2)+(i-1)*1+2;
    f(node11)=f(node11)+fel(1);
    f(node12)=f(node12)+fel(2);
    f(node21)=f(node21)+fel(3);
    f(node22)=f(node22)+fel(4);
end
K
```

```
K = 6×6
10⁸ ×
    0.3733    0.5600   -0.3733    0.5600         0         0
    0.5600    1.1200   -0.5600    0.5600         0         0
   -0.3733   -0.5600    0.7467         0   -0.3733    0.5600
    0.5600    0.5600         0    2.2400   -0.5600    0.5600
         0         0   -0.3733   -0.5600    0.3733   -0.5600
         0         0    0.5600    0.5600   -0.5600    1.1200
```

```
f
```

```
f = 6×1
     0
     0
     0
     0
     0
     0
```

The Assembled k matrix and the assembled f vector are shown above. The vector with all the Shear force and applied moments at the nodes have six members (two per nodal points).

$$\begin{bmatrix} V1 \\ M1 \\ V2 \\ M2 \\ V3 \\ M3 \end{bmatrix} \tag{4.23}$$

$$[K]\begin{bmatrix} u_1 \\ \theta_1 \\ u_2 \\ \theta_2 \\ u_3 \\ \theta_3 \end{bmatrix} = \begin{bmatrix} V1 \\ M1 \\ V2 \\ M2 \\ V3 \\ M3 \end{bmatrix} + [f] \tag{4.24}$$

The values at the nodes 1 and 3 are unknowns. The values at node 2 are known. So $V2 = -10{,}000$ N and $M2 = 20{,}000$ Nm. We use these known quantities and the elements in the 3rd and 4th row of the K matrix to solve for the unknown deflections and slopes first and then the unknown reaction forces and moments.

```
% commputing the unknown slopes and displacements from the reduced matrix
V = [V2;M2]

V = 2×1
        -10000
         20000

fsmall = [f(3);f(4)]

fsmall = 2×1
     0
     0

soln = K(3:4,3:4)\(V+fsmall)

soln = 2×1
10-3 ×
    -0.1339
     0.0893

% Computing All the shear and bending moment values at the nodes
disp=[0; 0; soln(1);soln(2);0 ;0]

disp = 6×1
10-3 ×

         0
         0
   -0.1339
    0.0893
         0
         0

V = K*disp - f

V = 6×1
        10000
        12500
       -10000
        20000
            0
        -2500
```

The shape functions of the beam are N1, N2, N3 and N4:

$$N_1 = 1 - \frac{3x^2}{L^2} + \frac{2x^3}{L^3}$$ (4.25)

$$N_2 = x - \frac{2x^2}{L} + \frac{x^3}{L^2}$$ (4.26)

$$N_3 = \frac{3x^2}{L^2} - \frac{2x^3}{L^3}$$ (4.27)

$$N_4 = -\frac{x^2}{L} + \frac{x^3}{L^2}$$ (4.28)

and beam deflection of a single element is given by:

$$u = u_1 N_1 + \theta_1 N_2 + u_2 N_3 + \theta_2 N_4$$ (4.29)

This relationship can be used to compute the deflection in each element and combining them for all the elements can enable the user to calculate the deflection of the entire beam (Fig. 4.3).

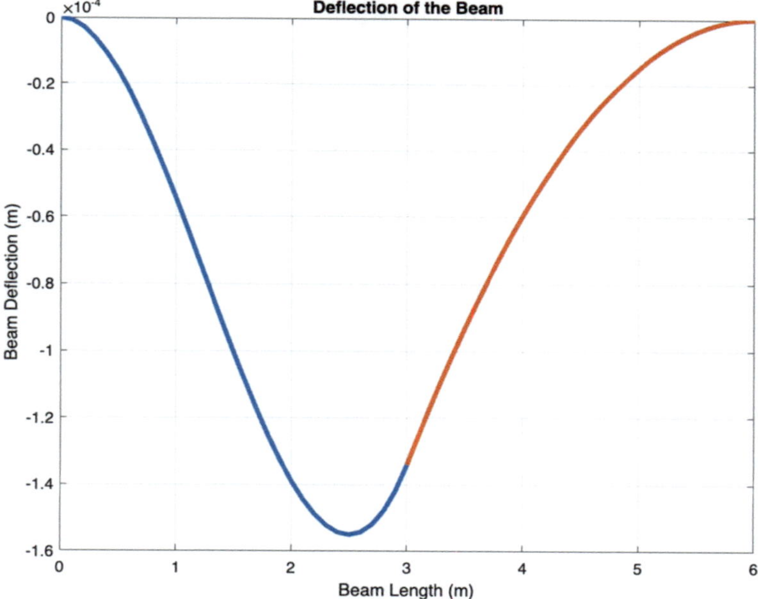

Fig. 4.3 Deflected curve representing the beam

```matlab
% Computing the  deflection across the beam and Drawing the deflected beam
figure
for i=1:elem
    %ke = k(i)*[1 -1;-1 1]
    node11 = elem_con(i,1)+(i-1)*1;
    node12 = elem_con(i,1)+(i-1)*1+1;
    node21 = elem_con(i,2)+(i-1)*1+1;
    node22 = elem_con(i,2)+(i-1)*1+2;

 y = (i-1)*L:0.1:i*L;
   x = y - (i-1)*L;
    discurve = disp(node11)*(1-3*x.^2/L^2+2*x.^3/L^3);
    discurve = discurve+disp(node12)*(x-2*x.^2/L+x.^3/L^2);
    discurve = discurve+ disp(node21)*(3*x.^2/L^2-2*x.^3/L^3);
    discurve = discurve+disp(node22)*(-x.^2/L+x.^3/L^2);

    plot(y,discurve,'linewidth',3);

hold   on

end

xlabel('Beam Length (m)');
ylabel('Beam Deflection (m)');
grid on
title('Deflection of the Beam');
```

Example 4.3 This is a beam that is on a roller at the left end and is fixed at the right end. It is 12 m long. At 4 m from the left end there is a concentrated force of 50,000 N acting downwards. E is 200 GPa and I is 80E−6 m^4.

In this case, we will divide the beam into three elements each of length 4 m. The end conditions will be, deflection is zero at the left end and deflection and slope are zero at the right end.

```matlab
clear
L = 4;
w = 0;
EI = 200E9*80E-6;
% Definnition of Stiffness Matrices
kel = EI/L^3*[12 6*L -12 6*L;6*L 4*L^2 -6*L 2*L^2;-12 -6*L 12 -6*L;
    6*L 2*L^2 -6*L 4*L^2]
```

```
kel = 4×4
10^7 ×
      0.3000      0.6000     -0.3000      0.6000
      0.6000      1.6000     -0.6000      0.8000
     -0.3000     -0.6000      0.3000     -0.6000
      0.6000      0.8000     -0.6000      1.6000
```

```matlab
fel = w*L/12*[6;L;6;-L];
% Applied Loads
M1 = 0
```

```
M1 = 0
```

```matlab
V2 = -50000
```

```
V2 = -50000
```

```matlab
M2 = 0
```

```
M2 = 0
```

```matlab
V3 = 0
```

```
V3 = 0
```

```matlab
M3 = 0
```

```
M3 = 0
```

```matlab
% Initializing
elem_con = [1 2;2 3;3 4]; %elemment connectivity
elem = size(elem_con,1);%total nummber of elements
nodes = elem+1;% in a 1 D system #nodes = #elements+1
K = zeros(2*nodes);%Initialize global stiffness matrix
f = zeros(2*nodes,1)% initialize the distributed load term
```

```
f = 8×1
     0
     0
     0
     0
     0
```

```matlab
       0
       0
       0

%Assembly
for i=1:elem
    %ke = k(i)*[1 -1;-1 1]
    node11 = elem_con(i,1)+(i-1)*1
    node12 = elem_con(i,1)+(i-1)*1+1
    node21 = elem_con(i,2)+(i-1)*1+1
    node22 = elem_con(i,2)+(i-1)*1+2
    K(node11,node11) = K(node11,node11) + kel(1,1);
    K(node11,node12) = K(node11,node12) + kel(1,2);
    K(node11,node21) = K(node11,node21) + kel(1,3);
    K(node11,node22) = K(node11,node22) + kel(1,4);
    %
    K(node12,node11) = K(node12,node11) + kel(2,1);
    K(node12,node12) = K(node12,node12) + kel(2,2);
    K(node12,node21) = K(node12,node21) + kel(2,3);
    K(node12,node22) = K(node12,node22) + kel(2,4);
    %
    K(node21,node11) = K(node21,node11) + kel(3,1);
    K(node21,node12) = K(node21,node12) + kel(3,2);
    K(node21,node21) = K(node21,node21) + kel(3,3);
    K(node21,node22) = K(node21,node22) + kel(3,4);
    %
    K(node22,node11) = K(node22,node11) + kel(4,1);
    K(node22,node12) = K(node22,node12) + kel(4,2);
    K(node22,node21) = K(node22,node21) + kel(4,3);
    K(node22,node22) = K(node22,node22) + kel(4,4);
end
    node11 = 1
    node12 = 2
    node21 = 3
    node22 = 4
    node11 = 3
    node12 = 4
    node21 = 5
    node22 = 6
    node11 = 5
    node12 = 6
    node21 = 7
    node22 = 8


% Create the Distributed load vector
```

```
for i=1:elem
    node11 = elem_con(i,1)+(i-1)*1;
    node12 = elem_con(i,1)+(i-1)*1+1;
    node21 = elem_con(i,2)+(i-1)*1+1;
    node22 = elem_con(i,2)+(i-1)*1+2;
    f(node11)=f(node11)+fel(1);
    f(node12)=f(node12)+fel(2);
    f(node21)=f(node21)+fel(3);
    f(node22)=f(node22)+fel(4);
end
K
```

```
K = 8×8
10⁷ ×
    0.3000    0.6000   -0.3000    0.6000        0         0         0 ...
    0.6000    1.6000   -0.6000    0.8000        0         0         0
   -0.3000   -0.6000    0.6000         0   -0.3000    0.6000         0
    0.6000    0.8000         0    3.2000   -0.6000    0.8000         0
         0         0   -0.3000   -0.6000    0.6000         0   -0.3000
         0         0    0.6000    0.8000         0    3.2000   -0.6000
         0         0         0         0   -0.3000   -0.6000    0.3000
         0         0         0         0    0.6000    0.8000   -0.6000
```

```
f
```

```
f = 8×1
    0
    0
    0
    0
    0
    0
    0
    0
```

The Assembled k matrix and the assembled f vector are shown here. The vector with all the Shear force and applied moments at the nodes have six members (two per nodal points.

$$\begin{bmatrix} V1 \\ M1 \\ V2 \\ M2 \\ V3 \\ M3 \\ V4 \\ M4 \end{bmatrix} \tag{4.30}$$

$$[K] \begin{bmatrix} u_1 \\ \theta_1 \\ u_2 \\ \theta_2 \\ u_3 \\ \theta_3 \\ u_4 \\ \theta_4 \end{bmatrix} = \begin{bmatrix} V1 \\ M1 \\ V2 \\ M2 \\ V3 \\ M3 \\ V4 \\ M4 \end{bmatrix} + [f] \tag{4.31}$$

The reaction forces at nodes 1 and 4 is unknowns. The reaction moment at node 1 is zero but at node 4 is unknown. The values at node 2,3 are known. So V2 = $-$ 50,000 N, M2 = 0, V3 = 0, M3 = 0. We use these known quantities and the elements in the 3rd and 4th row of the K matrix to solve for the unknown deflections and slopes first and then the unknown reaction forces and moments.

```
% commputing the unknown slopes and displacements from the reduced matrix
V = [M1;V2;M2;V3;M3]

V = 5×1
             0
        -50000
             0
             0
             0

fsmall = [f(2);f(3);f(4);f(5);f(6)]

fsmall = 5×1
         0
         0
         0
         0
         0

soln = K(2:6,2:6)\(V+fsmall)
```

```
soln = 5×1
   -0.0167
   -0.0494
   -0.0037
   -0.0284
    0.0102
```

```
% Computing All the shear and bending moment values at the nodes
disp=[0; soln(1); soln(2);soln(3);soln(4);soln(5);0;0]
```

```
disp = 8×1
         0
   -0.0167
   -0.0494
   -0.0037
   -0.0284
    0.0102
         0
         0
```

```
V = K*disp - f
```

```
V = 8×1
10⁴ ×
    2.5926
    0.0000
   -5.0000
   -0.0000
         0
    0.0000
    2.4074
   -8.8889
```

The shape functions of the beam are N1, N2, N3 and N4:

$$N_1 = 1 - \frac{3x^2}{L^2} + \frac{2x^3}{L^3} \tag{4.32}$$

$$N_2 = x - \frac{2x^2}{L} + \frac{x^3}{L^2} \tag{4.33}$$

$$N_3 = \frac{3x^2}{L^2} - \frac{2x^3}{L^3} \tag{4.34}$$

$$N_4 = -\frac{x^2}{L} + \frac{x^3}{L^2} \tag{4.35}$$

and beam deflection of a single element is given by:

$$u = u_1 N_1 + \theta_1 N_2 + u_2 N_3 + \theta_2 N_4 \tag{4.36}$$

This relationship can be used to compute the deflection in each element and combining them for all the elements can enable the user to calculate the deflection of the entire beam (Fig. 4.4).

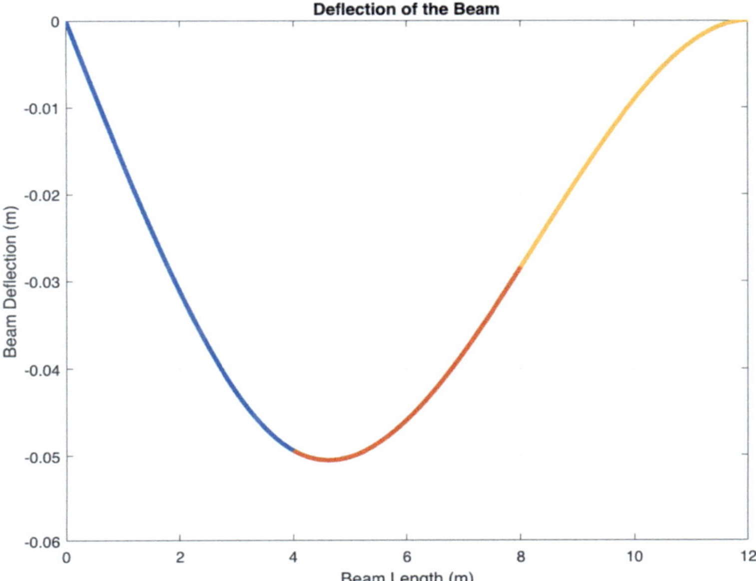

Fig. 4.4 Deflected shape of the beam

```matlab
% Computing the  deflection across the beam and Drawing the deflected beam
figure
for i=1:elem
    %ke = k(i)*[1 -1;-1 1]
    node11 = elem_con(i,1)+(i-1)*1;
    node12 = elem_con(i,1)+(i-1)*1+1;
    node21 = elem_con(i,2)+(i-1)*1+1;
    node22 = elem_con(i,2)+(i-1)*1+2;

  y = (i-1)*L:0.1:i*L;
   x = y - (i-1)*L;
    discurve = disp(node11)*(1-3*x.^2/L^2+2*x.^3/L^3);
    discurve = discurve+disp(node12)*(x-2*x.^2/L+x.^3/L^2);
    discurve = discurve+ disp(node21)*(3*x.^2/L^2-2*x.^3/L^3);
    discurve = discurve+disp(node22)*(-x.^2/L+x.^3/L^2);

    plot(y,discurve,'linewidth',3)

hold   on

end

xlabel('Beam Length (m)');
ylabel('Beam Deflection (m)');
grid on
title('Deflection of the Beam');
```

Example 4.4 See Fig. 4.5. In this problem we will use the PDE toolbox to setup the beam problem and then automatically solve it using MATLAB commands.

```
cantileverbeam = createpde('structural','static-planestress'); %model
```

Create a geometry with three adjacent rectangles. The top edge of the second rectangle (on the right) represents the location of the distributed load.

```
R1 = [3,4, [ 0, 1,  1,   0, -2, -2, 0, 0]/100]'; %definition of the first
region
R2 = [3,4, [1, 6, 6, 1, -2, -2, 0, 0]/100]';%defination of the second region
R3 = [3,4, [6, 12, 12, 6, -2,-2, 0, 0]/100]';%definition of the third region

gdm = [R1 R2 R3];
ns = char('R1','R2','R3');
g = decsg(gdm,'R1 + R2 + R3',ns'); %geometry definition as a combination of the
three regions
```

Assign the geometry to the model.

```
geometryFromEdges(cantileverbeam,g);
```

Plot the geometry with the edge, face labels and vertex labels (Fig. 4.6).

Fig. 4.5 Beam with distributed load on part of the beam

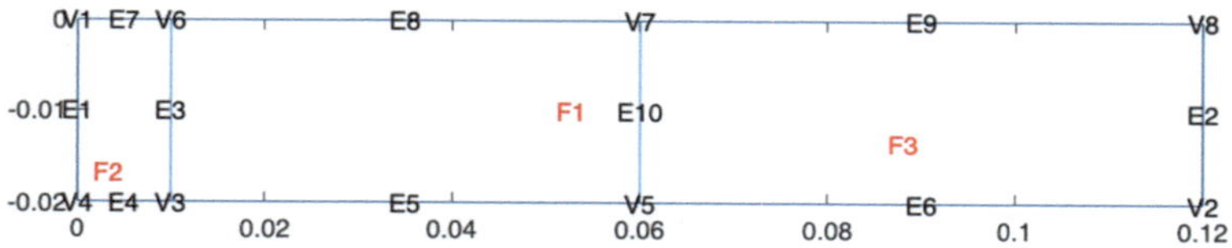

Fig. 4.6 Composite geometry for Example 4.4

```
figure
pdegplot(cantileverbeam,'EdgeLabels','on','FaceLabels','on','VertexLabels','on')
```

Generate a mesh. Use the linear geometric order instead of the default quadratic order (Fig. 4.7).

```
generateMesh(cantileverbeam,'Hmax',0.3E-02,'GeometricOrder','linear'); %
generate linear triangular elements
figure
pdeplot(cantileverbeam,"Mesh","on") %plot the mesh
axis equal
```

Specify Model Parameters

Specify the Young's modulus and Poisson's ratio to model linear elastic material behavior. Remember to specify physical properties in consistent units.

```
structuralProperties(cantileverbeam,'YoungsModulus',200E9,'PoissonsRatio',0.25);
```

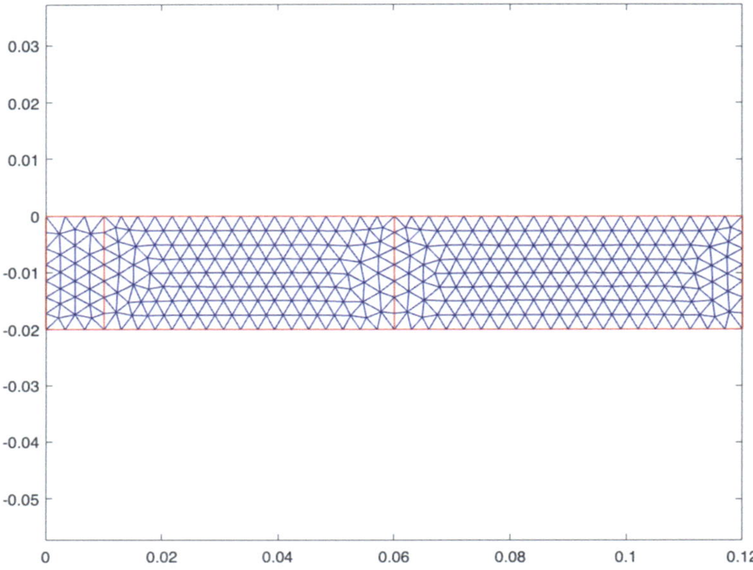

Fig. 4.7 Meshed geometry for Example 4.4

Restrain all rigid-body motions of the plate by specifying sufficient constraints. For static analysis, the constraints must also resist the motion induced by applied load.

In this case the left edge is fixed or cantilevered.

```
structuralBC(cantileverbeam,'Edge',1,'Constraint','fixed');
structuralBoundaryLoad(cantileverbeam,'Edge',8,'SurfaceTraction',[0;-1000]);
```

```
R = solve(cantileverbeam); % the solution is stored in the variable R
```

Plot simulation results. Figures 4.8, 4.9 and 4.10 show normal stress in x-direction, the deflected beam and the deflection contours across the beam, respectively.

```
figure
pdeplot(cantileverbeam,'XYData',R.Stress.xx,'ColorMap','jet');%plot the X-
stress
axis equal
title 'Normal Stress Along x-Direction';
```

```
figure
pdeplot(cantileverbeam,"Deformation",R.Displacement)%plot the deeformed
geometry
```

```
pdeplot(cantileverbeam,"XYData",R.Displacement.uy,"ColorMap",'jet') % plot the
vertical displacement
axis equal
title 'Vertical Beam Displacement';
```

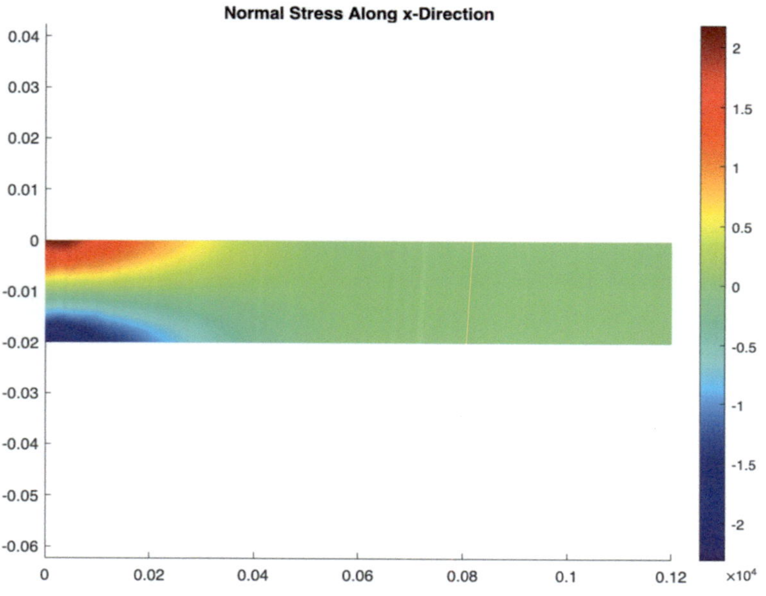

Fig. 4.8 Normal stress in x-direction

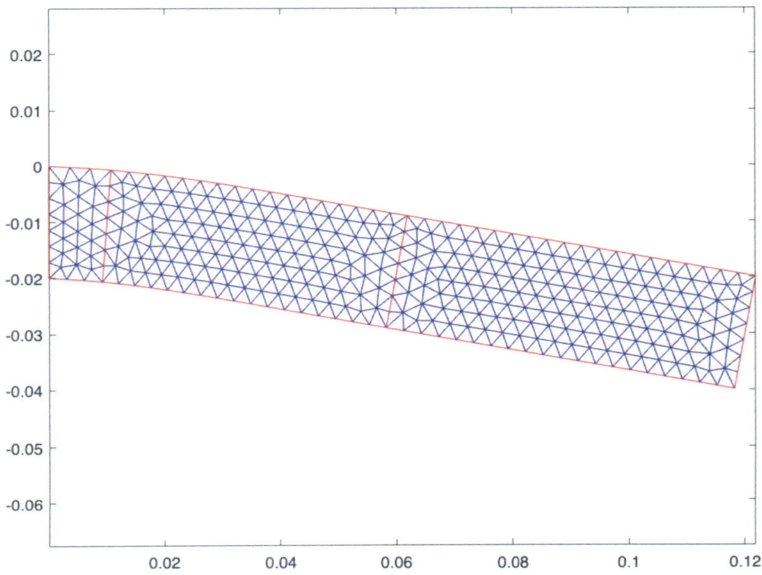

Fig. 4.9 Deflected beam

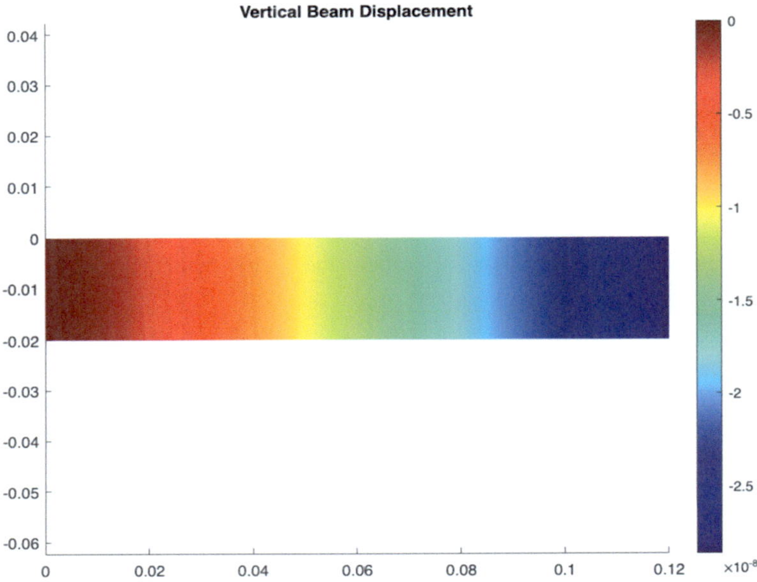

Fig. 4.10 Vertical displacement of the beam

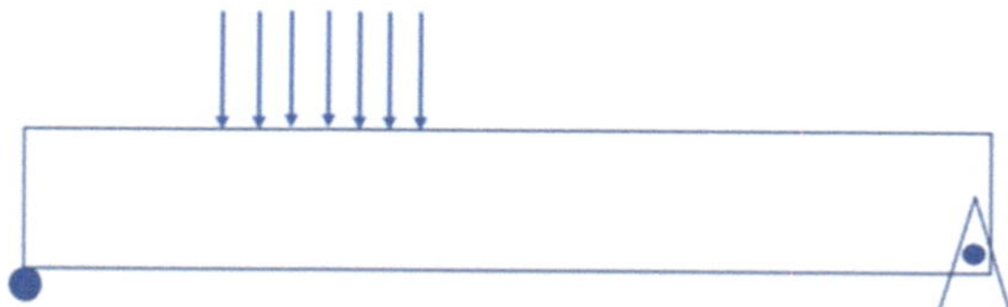

Fig. 4.11 Beam with simple supports

Example 4.5 Rerunning the model with simply supported boundary condition (Fig. 4.11)

```
simplebeam = createpde('structural','static-planestress'); %model
```

Create a geometry with two adjacent rectangles. The top edge of the longer rectangle (on the right) represents the disc-pad contact region.

```
R1 = [3,4, [ 0, 1,  1,   0, -2, -2, 0, 0]/100]'; %definition of the first
region
R2 = [3,4, [1, 6, 6, 1, -2, -2, 0, 0]/100]';%defination of the second region
R3 = [3,4, [6, 12, 12, 6, -2,-2, 0, 0]/100]';%definition of the third region

gdm = [R1 R2 R3];
ns = char('R1','R2','R3');
g = decsg(gdm,'R1 + R2 + R3',ns'); %geometry definition as a combination of the
three regions
```

Assign the geometry to the model.

```
geometryFromEdges(simplebeam,g);
```

Plot the geometry with the edge, face labels and vertex labels (Fig. 4.12).

```
figure
pdegplot(simplebeam,'EdgeLabels','on','FaceLabels','on','VertexLabels','on')
```

Generate a mesh. Use the linear geometric order instead of the default quadratic order (Fig. 4.13).

```
generateMesh(simplebeam,'Hmax',0.3E-02,'GeometricOrder','linear'); % generate
linear triangular elements
figure
pdeplot(simplebeam,"Mesh","on") %plot the mesh
axis equal
```

Specify Model Parameters.

Fig. 4.12 Composite geometry of Example 4.5

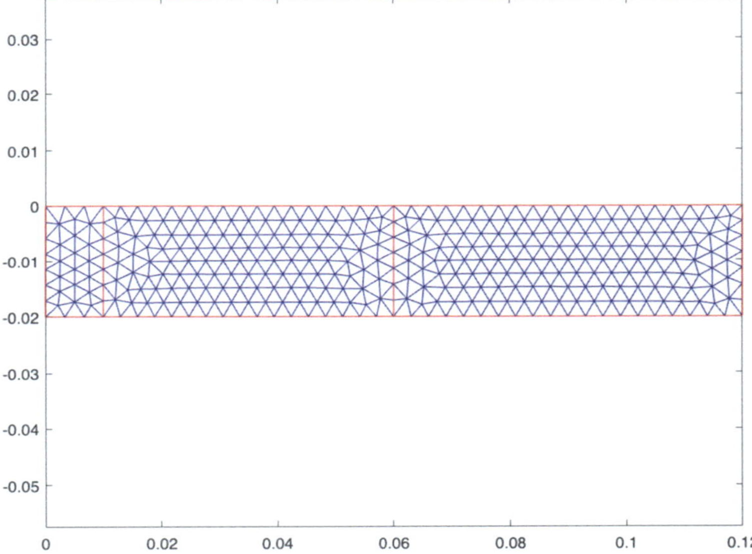

Fig. 4.13 Meshed geometry of Example 4.5

Specify the Young's modulus and Poisson's ratio to model linear elastic material behavior. Remember to specify physical properties in consistent units.

```
structuralProperties(simplebeam,'YoungsModulus',200E9,'PoissonsRatio',0.25);
```

Restrain all rigid-body motions of the plate by specifying sufficient constraints. For static analysis, the constraints must also resist the motion induced by applied load.

In this case the left edge is fixed or cantilevered.

```
structuralBC(simplebeam,'Vertex',2,'YDisplacement',0.0,'XDisplacement',0.0);
%structuralBC(simplebeam,'Vertex',2,'XDisplacement',0.0);
structuralBC(simplebeam,'Vertex',4,'YDisplacement',0.0);
structuralBoundaryLoad(simplebeam,'Edge',8,'SurfaceTraction',[0;-1000]);
```

```
R = solve(simplebeam); % the solution is stored in the variable R
```

Plot simulation results. Figures 4.14, 4.15 and 4.16 show normal stress in x-direction, the deflected beam and the deflection contours across the beam, respectively.

```
figure
pdeplot(simplebeam,'XYData',R.Stress.xx,'ColorMap','jet');%plot the X-stress
axis equal
title 'Normal Stress Along x-Direction';
```

```
figure
pdeplot(simplebeam,"Deformation",R.Displacement)%plot the deeformed geometry
```

```
pdeplot(simplebeam,"XYData",R.Displacement.uy,"ColorMap",'jet') % plot the
vertical displacement
axis equal
title 'Vertical Beam Displacement';
```

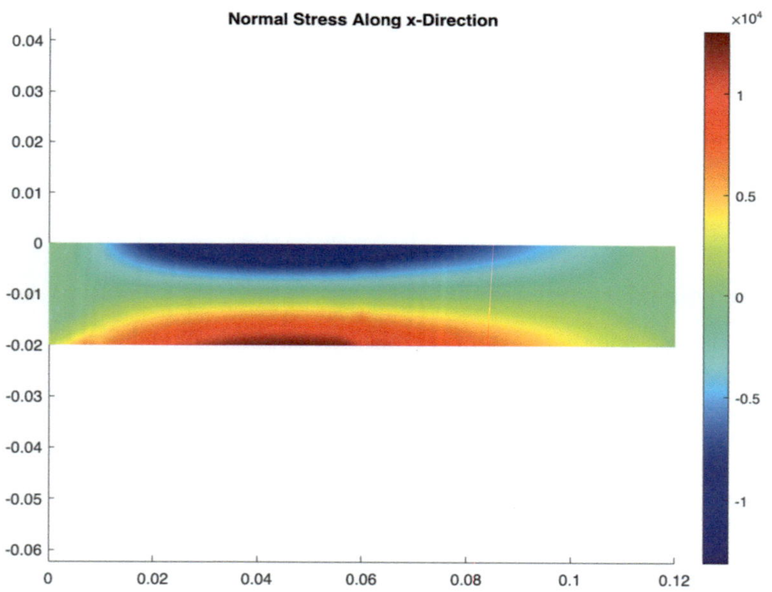

Fig. 4.14 Normal stress in X-direction

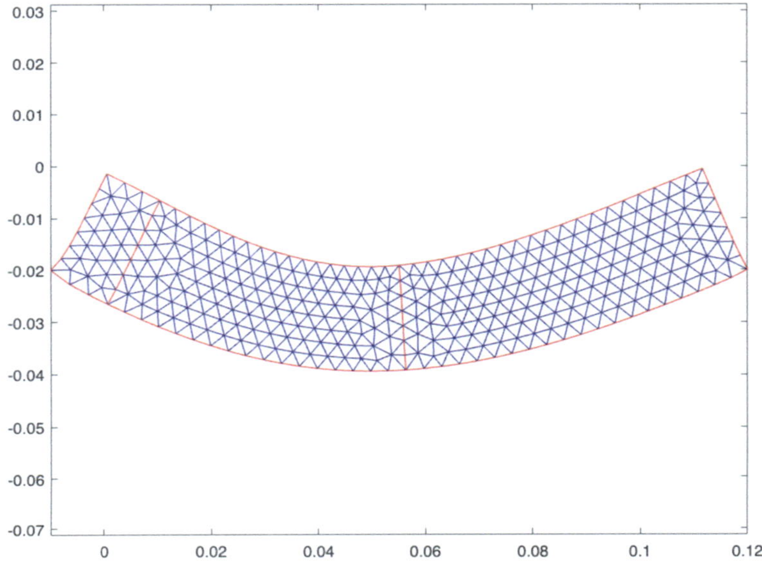

Fig. 4.15 Deflected beam

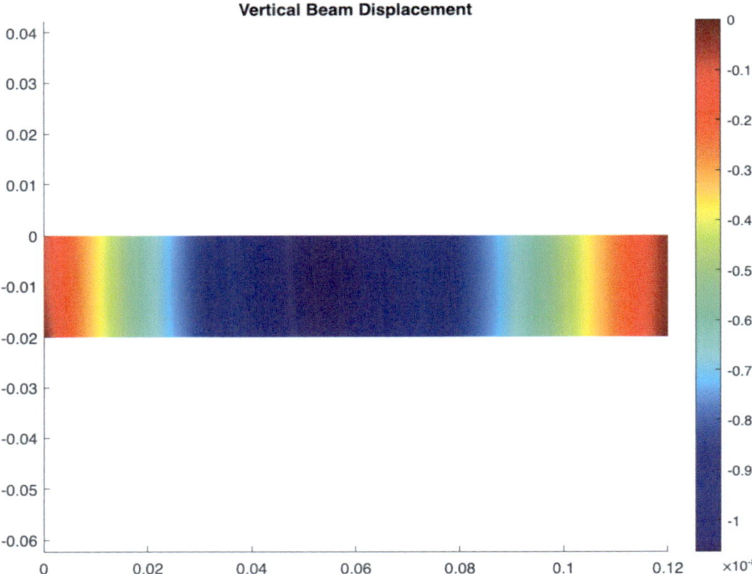

Fig. 4.16 Vertical displacement

4.2 Elastic Stress Analysis

Background information on Elasticity

The most general state of stress at a point is given by nine components of stress that are usually shown on three mutually perpendicular planes with normals pointing to three mutually perpendicular directions (Fig. 4.17).

The stress components are shown as stress tensor as:

$$S = \begin{bmatrix} \sigma_{xx} & \sigma_{xy} & \sigma_{xz} \\ \sigma_{yx} & \sigma_{yy} & \sigma_{yz} \\ \sigma_{zx} & \sigma_{zy} & \sigma_{zz} \end{bmatrix} = \begin{bmatrix} \sigma_{11} & \sigma_{12} & \sigma_{13} \\ \sigma_{21} & \sigma_{22} & \sigma_{23} \\ \sigma_{31} & \sigma_{32} & \sigma_{33} \end{bmatrix} \tag{4.37}$$

This is a symmetric matrix with the diagonal term representing the three normal stresses and the off-diagonal terms are the shear stresses. Since the matrix is symmetric, this means that shear stress $\sigma_{12} = \sigma_{21}$, $\sigma_{23} = \sigma_{32}$, etc. The stress tensor can be represented in the form of a stress vector as (using just the six components):

$$\begin{bmatrix} \sigma_{11} \\ \sigma_{22} \\ \sigma_{33} \\ \sigma_{12} \\ \sigma_{23} \\ \sigma_{31} \end{bmatrix} \tag{4.38}$$

Stresses on an Inclined plane

Stresses on an inclined plane or a plane that is at an angle to the three perpendicular planes can be obtained using a simple multiplication with the tensor. If the unit normal vector to the plane is given by $\begin{bmatrix} n_1 & n_2 & n_3 \end{bmatrix} = \begin{bmatrix} n_x & n_y & n_z \end{bmatrix}$ then the stress components on that plane are given by

$$t_n = (S)(n) \tag{4.39}$$

As an example let

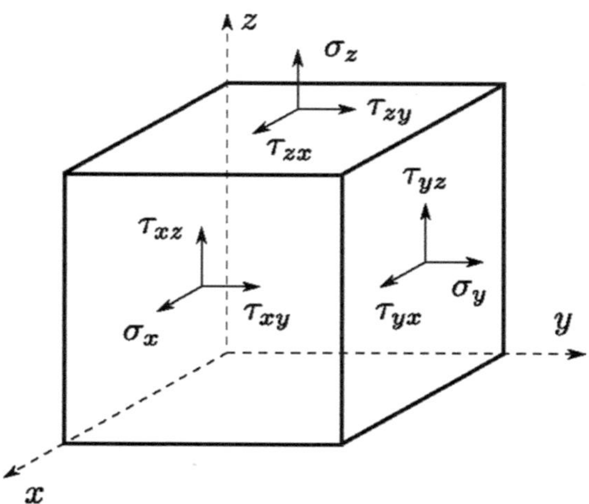

Fig. 4.17 Schematic showing generalized state of stress

$$S = \begin{bmatrix} 9 & -2 & 0 \\ -2 & -3 & 2 \\ 0 & 2 & 3 \end{bmatrix} \tag{4.40}$$

and let

$$n = \begin{bmatrix} \frac{2}{3} \\ \frac{1}{3} \\ \frac{2}{3} \end{bmatrix} \tag{4.41}$$

So the stress components on that plane can be computed as:

$$t_n = \begin{bmatrix} 9 & -2 & 0 \\ -2 & -3 & 2 \\ 0 & 2 & 3 \end{bmatrix} \begin{bmatrix} \frac{2}{3} \\ \frac{1}{3} \\ \frac{2}{3} \end{bmatrix} = \begin{bmatrix} 5.333 \\ -1 \\ 0.667 \end{bmatrix} \tag{4.42}$$

From this data the normal and shear stresses can be computed by taking its component along the normal direction:

$$\sigma_n = n^T t_n \tag{4.43}$$

and the shear stress can be computed as:

$$\tau_n = \sqrt{\left(t_n^T t_n - \sigma_n^2\right)} \tag{4.44}$$

```
S =[9 -2 0;-2 -3 2;0 2 3]

S = 3×3
       9      -2       0
      -2      -3       2
       0       2       3

n = [2/3;1/3;2/3]

n = 3×1
    0.6667
    0.3333
    0.6667

t = S*n

t = 3×1
    5.3333
   -1.0000
    2.6667

sigman = (n)'*t

sigman = 5

shear = sqrt(t'*t-sigman^2)

shear = 3.3993
```

Finding Principal Stresses and planes for Principal stresses

Principal stresses are eigen values and their directions are eigen vectors for the Stress matrix.

```
 S =[10 -3 0;-3 -7 2;0 2 5]

S = 3×3
      10     -3      0
      -3     -7      2
       0      2      5

 [V,D] = eig(S)

V = 3×3
     0.1641    0.0884    0.9825
     0.9746    0.1390   -0.1753
    -0.1521    0.9863   -0.0633
D = 3×3
    -7.8172         0         0
          0    5.2819         0
          0         0   10.5353
```

So, in this case the V matrix represents the three directions and the D matrix diagonal elements represent the three principal stresses.

Failure Criteria

There are two well-known failure criteria that are commonly used in practice, the Tresca and the Von Mises criteria.

Tresca

The Tresca stress/criteria is based on the maximum shear stress theory. The maximum shear stress at a point is given by:

$\tau_{\max} = \frac{\sigma_1 - \sigma_3}{2}$, where the two normal stresses are the highest and the smallest principal stresses at the point.

In this failure criteria the maximum shear stress is compared with the maximum shear stress in the uniaxial loading condition.

So, as per the Tresca criterion, failure will occur if:

$$\tau_{\max} = \frac{\sigma_1 - \sigma_3}{2} \geq \frac{\sigma_Y}{2} \tag{4.45}$$

and the factor of safety in design will be:

$$\mathrm{FOS} = \frac{\sigma_Y}{(\sigma_1 - \sigma_3)} \tag{4.46}$$

Von Misses

The Von Mises criteria is based on the maximum strain energy theory. The Von Mises stress is written as:

$$\sigma_v = \sqrt{\frac{(\sigma_{11} - \sigma_{22})^2 + (\sigma_{22} - \sigma_{33})^2 + (\sigma_{33} - \sigma_{11})^2 + 6(\sigma_{12}^2 + \sigma_{23}^2 + \sigma_{31}^2)}{2}}$$

$$= \sqrt{\frac{(\sigma_1 - \sigma_2)^2 + (\sigma_2 - \sigma_3)^2 + (\sigma_3 - \sigma_1)^2}{2}} \qquad (4.47)$$

As per this failure criteria: $\sigma_V \geq \sigma_Y$ will cause failure.

Strain Displacement Relationships

The most general state of strain at a point is given by nine components of strain that are usually shown on three mutually perpendicular planes with normals pointing to three mutually perpendicular directions. The strain components are shown as strain tensor as:

$$\begin{bmatrix} \epsilon_{xx} & \epsilon_{xy} & \epsilon_{xz} \\ \epsilon_{yx} & \epsilon_{yy} & \epsilon_{yz} \\ \epsilon_{zx} & \epsilon_{zy} & \epsilon_{zz} \end{bmatrix} = \begin{bmatrix} \epsilon_{11} & \epsilon_{12} & \epsilon_{13} \\ \epsilon_{21} & \epsilon_{22} & \epsilon_{23} \\ \epsilon_{31} & \epsilon_{32} & \epsilon_{33} \end{bmatrix} \qquad (4.48)$$

This is a symmetric matrix with the diagonal term representing the three normal strains and the off-diagonal terms are the shear strains. Since the matrix is symmetric, this means that shear strains $\epsilon_{12} = \epsilon_{21}$, $\epsilon_{23} = \epsilon_{32}$, etc.

Strain is related to displacement or deformation. If the deformation in the three directions is given by u_1, u_2, u_3, strain can be defined as:

$$\epsilon_{ij} = \frac{1}{2}\left(\frac{\partial u_i}{\partial x_j} + \frac{\partial u_j}{\partial x_i}\right) \qquad (4.49)$$

This means when i and j are equal to each other, e.g. 1, 2 or 3 we get the three normal strains as:

$$\epsilon_{11} = \frac{1}{2}\left(\frac{\partial u_1}{\partial x_1} + \frac{\partial u_1}{\partial x_1}\right) = \frac{\partial u_1}{\partial x_1} \qquad (4.50)$$

$$\epsilon_{22} = \frac{1}{2}\left(\frac{\partial u_2}{\partial x_2} + \frac{\partial u_2}{\partial x_2}\right) = \frac{\partial u_2}{\partial x_2} \qquad (4.51)$$

$$\epsilon_{33} = \frac{1}{2}\left(\frac{\partial u_3}{\partial x_3} + \frac{\partial u_3}{\partial x_3}\right) = \frac{\partial u_3}{\partial x_3} \qquad (4.52)$$

When i and j are different, we get the shear strain components

$$\epsilon_{12} = \frac{1}{2}\left(\frac{\partial u_1}{\partial x_2} + \frac{\partial u_2}{\partial x_1}\right) \qquad (4.53)$$

$$\epsilon_{23} = \frac{1}{2}\left(\frac{\partial u_2}{\partial x_3} + \frac{\partial u_3}{\partial x_2}\right) \tag{4.54}$$

$$\epsilon_{31} = \frac{1}{2}\left(\frac{\partial u_3}{\partial x_1} + \frac{\partial u_1}{\partial x_3}\right) \tag{4.55}$$

Also, shear strain that is written as γ is relatable to the above definition of shear strain as:

$$\gamma_{ij} = 2\epsilon_{ij} \tag{4.56}$$

The strain tensor can also be represented in the form of a strain vector as (using just the six components) as:

$$\begin{bmatrix} \epsilon_{11} \\ \epsilon_{22} \\ \epsilon_{33} \\ \gamma_{12} \\ \gamma_{23} \\ \gamma_{31} \end{bmatrix} \tag{4.57}$$

Stresses and Strains are related to each other through the generalized stresss strain relationships as:

$$\begin{bmatrix} \sigma_{11} \\ \sigma_{22} \\ \sigma_{33} \\ \sigma_{12} \\ \sigma_{23} \\ \sigma_{31} \end{bmatrix} = \begin{bmatrix} C11 & C12 & C13 & C14 & C15 & C16 \\ C21 & C22 & C23 & C24 & C25 & C26 \\ C31 & C32 & C33 & C34 & C35 & C36 \\ C41 & C42 & C43 & C44 & C45 & C46 \\ C51 & C52 & C53 & C54 & C55 & C56 \\ C61 & C62 & C63 & C64 & C65 & C66 \end{bmatrix} \begin{bmatrix} \epsilon_{11} \\ \epsilon_{22} \\ \epsilon_{33} \\ \gamma_{12} \\ \gamma_{23} \\ \gamma_{31} \end{bmatrix} \tag{4.58}$$

The 36 coefficients have to be established for a material. This is the most generalized situation of non-homogeneous, non-isotropic material. Several practical considerations reduce the number of constants. For example, for linear elasticity the coefficient matrix is assumed to be symmetric. That reduces the number to 21. If the material is assumed to be symmetric about the three mutually perpendicular plane, then the number of constants reduce to 9

$$\begin{bmatrix} \sigma_{11} \\ \sigma_{22} \\ \sigma_{33} \\ \sigma_{12} \\ \sigma_{23} \\ \sigma_{31} \end{bmatrix} = \begin{bmatrix} C11 & C12 & C13 & 0 & 0 & 0 \\ C12 & C22 & C23 & 0 & 0 & 0 \\ C13 & C23 & C33 & 0 & 0 & 0 \\ 0 & 0 & 0 & C44 & 0 & 0 \\ 0 & 0 & 0 & 0 & C55 & 0 \\ 0 & 0 & 0 & 0 & 0 & C66 \end{bmatrix} \begin{bmatrix} \epsilon_{11} \\ \epsilon_{22} \\ \epsilon_{33} \\ \gamma_{12} \\ \gamma_{23} \\ \gamma_{31} \end{bmatrix} \tag{4.59}$$

Finally, if the material is assumed to be isotropic i.e., it is symmetric about every plane then the number of constants reduce to 2. They are Modulus of elasticity and Poisson's ratio.

$$\begin{bmatrix} \sigma_{11} \\ \sigma_{22} \\ \sigma_{33} \\ \sigma_{12} \\ \sigma_{23} \\ \sigma_{31} \end{bmatrix} = \frac{E}{(1+v)(1-2v)} \begin{bmatrix} 1-v & v & v & 0 & 0 & 0 \\ v & 1-v & v & 0 & 0 & 0 \\ v & v & 1-v & 0 & 0 & 0 \\ 0 & 0 & 0 & \frac{(1-2v)}{2} & 0 & 0 \\ 0 & 0 & 0 & 0 & \frac{(1-2v)}{2} & 0 \\ 0 & 0 & 0 & 0 & 0 & \frac{(1-2v)}{2} \end{bmatrix} \begin{bmatrix} \epsilon_{11} \\ \epsilon_{22} \\ \epsilon_{33} \\ \gamma_{12} \\ \gamma_{23} \\ \gamma_{31} \end{bmatrix}$$

$$(4.60)$$

The inverse of this representation is written as:

$$[\epsilon] = [D][\sigma] \tag{4.61}$$

where the D matrix is written as:

$$\begin{bmatrix} \frac{1}{E} & -\frac{v}{E} & -\frac{v}{E} & 0 & 0 & 0 \\ -\frac{v}{E} & \frac{1}{E} & -\frac{v}{E} & 0 & 0 & 0 \\ -\frac{v}{E} & -\frac{v}{E} & \frac{1}{E} & 0 & 0 & 0 \\ 0 & 0 & 0 & \frac{1}{G} & 0 & 0 \\ 0 & 0 & 0 & 0 & \frac{1}{G} & 0 \\ 0 & 0 & 0 & 0 & 0 & \frac{1}{G} \end{bmatrix} \tag{4.62}$$

where G, the modulus of rigidity, is

$$G = \frac{E}{2(1+v)} \tag{4.63}$$

Governing Differential Equations

The equilibrium equations for elastic solids is given by:

$$\frac{\partial \sigma_{xx}}{\partial x} + \frac{\partial \sigma_{xy}}{\partial y} + \frac{\partial \sigma_{xz}}{\partial z} + f_x = 0 \tag{4.64}$$

$$\frac{\partial \sigma_{yx}}{\partial x} + \frac{\partial \sigma_{yy}}{\partial y} + \frac{\partial \sigma_{yz}}{\partial z} + f_y = 0 \tag{4.65}$$

$$\frac{\partial \sigma_{zx}}{\partial x} + \frac{\partial \sigma_{zy}}{\partial y} + \frac{\partial \sigma_{zz}}{\partial z} + f_z = 0 \tag{4.66}$$

Typical boundary conditions in real problems are:

(a) Part of the boundary where the ith component of the traction is provided
(b) Part of the boundary where the ith component of displacement is provided
(c) Part of the boundary where a combined condition holds, i.e., the traction is a function of the displacement.

Plane Elasticity

There are several critical and practical simplifications that can be made to simplify a 3-D problem to a 2-D problem so that an effective analysis can be performed. Two of those are: Plane Stress condition and Plane Strain condition.

Plane Stress Condition

When all the off-pane stresses are equal to zero, i.e., $\sigma_{zz} = \sigma_{xz} = \sigma_{yz} = 0$ or $0 = \sigma_{33} = \sigma_{13} = \sigma_{23}$. This could occur when the loaded component is a thin section and the surfaces are accessible, i.e., they are plate-like structures. Off-plane strains are non-zero for plane-stress problems.

Plane Strain Condition

When all the off-pane strains are equal to zero, i.e., $\epsilon_{zz} = \epsilon_{xz} = \epsilon_{yz} = 0$ or $0 = \epsilon_{33} = \epsilon_{13} = \epsilon_{23}$. This could occur when the loaded component is a thick section and the surfaces are usually not accessible, e.g., mid-section section of a large dam. Off-plane stresses are non-zero for plane-strain problems.

The stress strain relationship changes accordingly

$$\begin{bmatrix} \sigma_{xx} \\ \sigma_{yy} \\ \sigma_{xy} \end{bmatrix} = \begin{bmatrix} C11 & C12 & 0 \\ C12 & C22 & 0 \\ 0 & 0 & C66 \end{bmatrix} \begin{bmatrix} \epsilon_{xx} \\ \epsilon_{yy} \\ \gamma_{xy} \end{bmatrix} \tag{4.67}$$

For plane stress:

$$C11 = C22 = \frac{E}{(1 - v^2)} \tag{4.68}$$

$$C12 = \frac{vE}{(1 - v^2)} \tag{4.69}$$

$$C66 = G = \frac{E}{2(1 + v)} \tag{4.70}$$

For plane strain:

$$C11 = C22 = \frac{E(1 - v)}{(1 - v - 2v^2)} \tag{4.71}$$

$$C12 = \frac{vE}{(1 - v - 2v^2)} \tag{4.72}$$

$$C66 = G = \frac{E}{2(1 + v)} \tag{4.73}$$

Boundary Conditions

Essential Boundary Conditions (Displacements are given)

$$\begin{aligned}
u_1 &= given\,value \\
u_2 &= given\,value
\end{aligned} \tag{4.74}$$

Natural Boundary Conditions (specified boundary tractions)

$$t_x = \sigma_{xx}n_x + \sigma_{xy}n_y \tag{4.75}$$

$$t_y = \sigma_{xy}n_x + \sigma_{yy}n_y \tag{4.76}$$

Weak Form for 2-D problems

In 2-D, the equilibrium equations become:

$$\frac{\partial \sigma_{xx}}{\partial x} + \frac{\partial \sigma_{xy}}{\partial y} + f_x = 0 \tag{4.77}$$

$$\frac{\partial \sigma_{yx}}{\partial x} + \frac{\partial \sigma_{yy}}{\partial y} + f_y = 0 \tag{4.78}$$

The two equations are multiplied by two test functions (virtual displacements)

$$(\bar{u}_1, \bar{u}_2) = (\bar{u}, \bar{v}) \tag{4.79}$$

We multiply the equations with the test function and write them out:

$$\int_V \sum_{i=1}^{2} \sum_{j=1}^{2} \left(-\frac{\partial \sigma_{ji}}{\partial x_j} \bar{u}_i \right) dV = \int_V \sum_{i=1}^{2} f_i \bar{u}_i dV \tag{4.80}$$

Recall the Green-Gauss-Divergence Theorem

$$\int_V \beta \nabla^2 G dV = \int_S \beta \nabla G . \hat{n}\, dS - \int_V \nabla G \nabla \beta\, dA \tag{4.81}$$

Let,

$$\nabla G = \sigma, \beta = \bar{u} \tag{4.82}$$

$$\int_V \bar{u}\nabla\sigma\,\mathrm{dV} = \int_S \bar{u}\sigma \cdot \hat{n}\,\mathrm{dS} - \int_V \sigma\nabla\bar{u}\,\mathrm{dV} \tag{4.83}$$

Substituting this in the integral equation we get:

$$\sum_{i=1}^{2}\sum_{j=1}^{2}\int_V \sigma_{ji}\frac{\partial\bar{u}_i}{\partial x_j}\,\mathrm{dV} = \sum_{i=1}^{2}\int_S \sigma_{ji}\bar{u}_{ii}n_j\,\mathrm{dS} + \sum_{i=1}^{2}\int_V f_i\bar{u}_i\,\mathrm{dV} \tag{4.84}$$

$$\sum_{i=1}^{2}\sum_{j=1}^{2}\int_V \sigma_{ji}\frac{\partial\bar{u}_i}{\partial x_j}\,\mathrm{dV} = \sum_{i=1}^{2}\int_S t_i\bar{u}_i\,\mathrm{dS} + \sum_{i=1}^{2}\int_V f_i\bar{u}_i\,\mathrm{dV} \tag{4.85}$$

Consider the first term in Eq. 4.85. With t being uniform thickness the volume integral becomes an area integral and a surface integral has become a line integral.

$$\sum_{i=1}^{2}\sum_{j=1}^{2}\int_V \sigma_{ji}\frac{\partial\bar{u}_i}{\partial x_j}\,\mathrm{dV} = \sum_{i=1}^{2}\sum_{j=1}^{2} t\int \sigma_{ji}\frac{\partial\bar{u}_i}{\partial x_j}\,\mathrm{dA} \tag{4.86}$$

$$\sum_{i=1}^{2}\sum_{j=1}^{2} t\int_A \sigma_{ji}\frac{\partial\bar{u}_i}{\partial x_j}\,\mathrm{dA} = \sum_{i=1}^{2}\sum_{j=1}^{2} t\int_A \frac{1}{2}\left(\sigma_{ij} + \sigma_{ji}\right)\frac{\partial\bar{u}_i}{\partial x_j}\,\mathrm{dA} \tag{4.87}$$

$$\sum_{i=}^{2}\sum_{j=1}^{2}\frac{1}{2}\left(\sigma_{ij} + \sigma_{ji}\right)\frac{\partial\bar{u}_i}{\partial x_j} = \sum_{i=1}^{2}\sum_{j=1}^{2}\sigma_{ji}\frac{1}{2}\left(\frac{\partial\bar{u}_j}{\partial x_i} + \frac{\partial\bar{u}_i}{\partial x_j}\right) \tag{4.88}$$

Expanding the summation, we get:

$$\sigma_{11}\frac{\partial\bar{u}_1}{\partial x_1} + \sigma_{12}\left(\frac{\partial\bar{u}_2}{\partial x_1} + \frac{\partial\bar{u}_1}{\partial x_2}\right) + \sigma_{22}\frac{\partial\bar{u}_2}{\partial x_2} = \epsilon^T\sigma \tag{4.89}$$

so the integral equation becomes (there is a volume integral that still remains which will also get converted to an area integral times thickness):

$$t\int_A \epsilon^T\sigma\,\mathrm{dA} = \sum_{i=1}^{2}\int_S t_i\bar{u}_i\,\mathrm{dS} + \sum_{i=1}^{2}\int_V f_i\bar{u}_i\,\mathrm{dV} \tag{4.90}$$

Fig. 4.18 3-noded triangular element with displacements u1 and u2 at each node

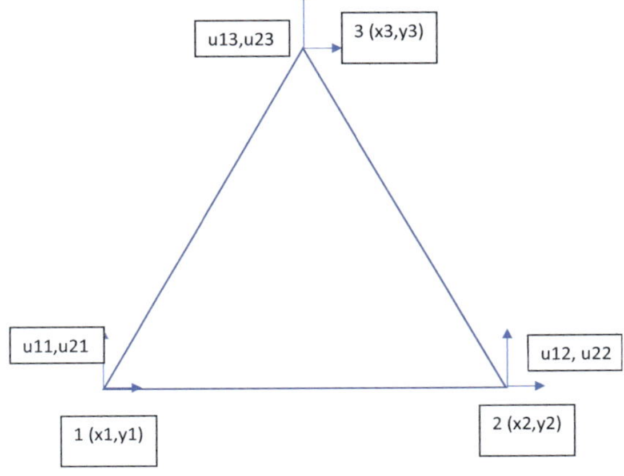

$$t \int_A \left[\epsilon^T \right][D][\epsilon]\mathrm{dA} = \sum_{i=1}^{2} \int_S t_i \bar{u}_i \mathrm{dS} + \sum_{i=1}^{2} \int_V f_i \bar{u}_i \mathrm{dV} \tag{4.91}$$

Triangular Element (Fig. 4.18)

Let us consider a 3-noded triangular element with the nodes as i, j and k the displacements in the two directions can be written as:

$$(u, v) = (u_1, u_2) \tag{4.92}$$

So,

$$u_1 = u_{1i}N_i + u_{1j}N_j + u_{1k}N_k \tag{4.93}$$

$$u_2 = u_{2i}N_i + u_{2j}N_j + u_{2k}N_k \tag{4.94}$$

If we define [N] as:

$$[N] = \begin{bmatrix} N_i & 0 & N_j & 0 & N_k & 0 \\ 0 & N_i & 0 & N_j & 0 & N_k \end{bmatrix} \tag{4.95}$$

So the displacement vector can be written as:

$$\begin{bmatrix} u_1 \\ u_2 \end{bmatrix} = [N] \begin{bmatrix} u_{1i} \\ u_{2i} \\ u_{1j} \\ u_{2j} \\ u_{3k} \\ u_{3k} \end{bmatrix} \tag{4.96}$$

The strain matrix can be written as:

$$\begin{bmatrix} \epsilon_{11} \\ \epsilon_{22} \\ \gamma_{12} \end{bmatrix} = \begin{bmatrix} \frac{\partial N_i}{\partial x} & 0 & \frac{\partial N_j}{\partial x} & 0 & \frac{\partial N_k}{\partial x} & 0 \\ 0 & \frac{\partial N_i}{\partial y} & 0 & \frac{\partial N_j}{\partial y} & 0 & \frac{\partial N_j}{\partial y} \\ \frac{\partial N_i}{\partial y} & \frac{\partial N_i}{\partial x} & \frac{\partial N_j}{\partial y} & \frac{\partial N_j}{\partial x} & \frac{\partial N_k}{\partial y} & \frac{\partial N_k}{\partial x} \end{bmatrix} \begin{bmatrix} u_{1i} \\ u_{2i} \\ u_{1j} \\ u_{2j} \\ u_{1k} \\ u_{2k} \end{bmatrix} \tag{4.97}$$

Since strains are defined in the following way:

$$\epsilon_{11} = \frac{1}{2}\left(\frac{\partial u_1}{\partial x_1} + \frac{\partial u_1}{\partial x_1}\right) = \frac{\partial u_1}{\partial x_1} \tag{4.98}$$

$$\epsilon_{22} = \frac{1}{2}\left(\frac{\partial u_2}{\partial x_2} + \frac{\partial u_2}{\partial x_2}\right) = \frac{\partial u_2}{\partial x_2} \tag{4.99}$$

$$\epsilon_{33} = \frac{1}{2}\left(\frac{\partial u_3}{\partial x_3} + \frac{\partial u_3}{\partial x_3}\right) = \frac{\partial u_3}{\partial x_3} \tag{4.100}$$

When i and j are different, we get the shear stress components

$$\epsilon_{12} = \frac{1}{2}\left(\frac{\partial u_1}{\partial x_2} + \frac{\partial u_2}{\partial x_1}\right) \tag{4.101}$$

$$\epsilon_{23} = \frac{1}{2}\left(\frac{\partial u_2}{\partial x_3} + \frac{\partial u_3}{\partial x_2}\right) \tag{4.102}$$

$$\epsilon_{31} = \frac{1}{2}\left(\frac{\partial u_3}{\partial x_1} + \frac{\partial u_1}{\partial x_3}\right) \tag{4.103}$$

and

$$\gamma_{ij} = 2\epsilon_{ij} \tag{4.104}$$

$$[B] = \begin{bmatrix} \frac{\partial N_i}{\partial x} & 0 & \frac{\partial N_j}{\partial x} & 0 & \frac{\partial N_k}{\partial x} & 0 \\ 0 & \frac{\partial N_i}{\partial y} & 0 & \frac{\partial N_j}{\partial y} & 0 & \frac{\partial N_j}{\partial y} \\ \frac{\partial N_i}{\partial y} & \frac{\partial N_i}{\partial x} & \frac{\partial N_j}{\partial y} & \frac{\partial N_j}{\partial x} & \frac{\partial N_k}{\partial y} & \frac{\partial N_k}{\partial x} \end{bmatrix} \tag{4.105}$$

So,

$$[\epsilon] = [B][u] \tag{4.106}$$

and the left hand side of Eq. 4.91 can be written as,

$$t \int_A [\epsilon^T][D][\epsilon]\mathrm{dA} = t \int_A [\bar{u}]^T[B]^T[D][B][u]\mathrm{dA} = [\bar{u}]^T tA[B]^T[D][B][u] \quad (4.107)$$

$$[\bar{u}]^T tA[B]^T[D][B][u] = [\bar{u}]^T[K][u] \quad (4.108)$$

where t is uniform thickness of the element.

Consider the 2nd term on the right-hand side of 4.85

$$\sum_{i=1}^{2} \int_V f_i \bar{u}_i \mathrm{dV} = \int_V [\bar{u}_i]^T[f_i]\mathrm{dV} = [\bar{u}]^T \int_V [N]^T[f_i]\mathrm{dV} \quad (4.109)$$

$$[\bar{u}]^T \int_V [N]^T[f_i]\mathrm{dv} = [\bar{u}]^T \int_V \begin{bmatrix} f_x & N_i \\ f_y & N_i \\ f_x & N_j \\ f_y & N_j \\ f_x & N_k \\ f_y & N_k \end{bmatrix} \mathrm{dV}$$

$$= [\bar{u}]^T \frac{tA}{3} \begin{bmatrix} f_x \\ f_y \\ f_x \\ f_y \\ f_x \\ f_y \end{bmatrix} = [\bar{u}]^T[f^1] \quad (4.110)$$

Consider the 1st term on the right-hand side of 4.86.

This is a boundary integral which means that this term is applicable only for elements that have a side on the boundary. For triangular elements only one of the three sides could be on the boundary (Fig. 4.19).

Case 1 Side i-j on the boundary

$$\sum_{i=1}^{2} \int_S t_i \bar{u}_i \mathrm{dS} = [\bar{u}]^T \begin{bmatrix} \frac{t_x L}{2} \\ \frac{t_y L}{2} \\ \frac{t_x L}{2} \\ \frac{t_y L}{2} \\ 0 \\ 0 \end{bmatrix} \quad (4.111)$$

Fig. 4.19 Elements on the boundary with different node combinations

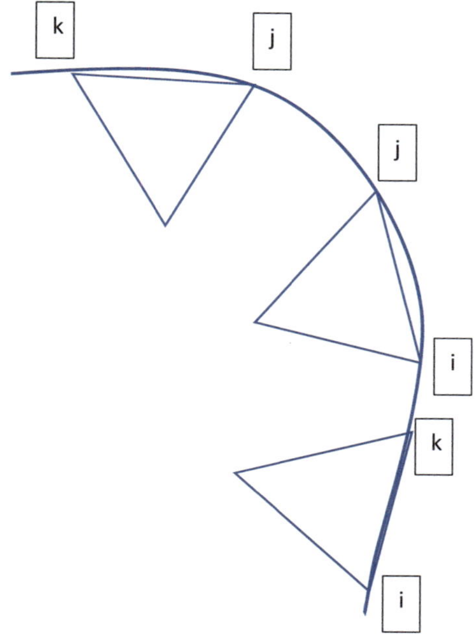

Case 2 Side j-k on the boundary

$$\sum_{i=1}^{2} \int_{S} t_i \, \overline{u}_{\,i} \mathrm{dS} = \left[\,\overline{u}\,\right]^{T} \begin{bmatrix} 0 \\ 0 \\ \frac{t_x L}{2} \\ \frac{t_y L}{2} \\ \frac{t_x L}{2} \\ \frac{t_y L}{2} \end{bmatrix} \tag{4.112}$$

Case 2 Side k-i on the boundary

$$\sum_{i=1}^{2} \int_{S} t_i \overline{u}_i \mathrm{dS} = \left[\,\overline{u}\,\right]^{T} \begin{bmatrix} \frac{t_x L}{2} \\ \frac{t_y L}{2} \\ 0 \\ 0 \\ \frac{t_x L}{2} \\ \frac{t_y L}{2} \end{bmatrix} \tag{4.113}$$

We will refer to this boundary integral term as: $\left[\,\overline{u}\,\right]^{T}\left[f^{2}\right]$.

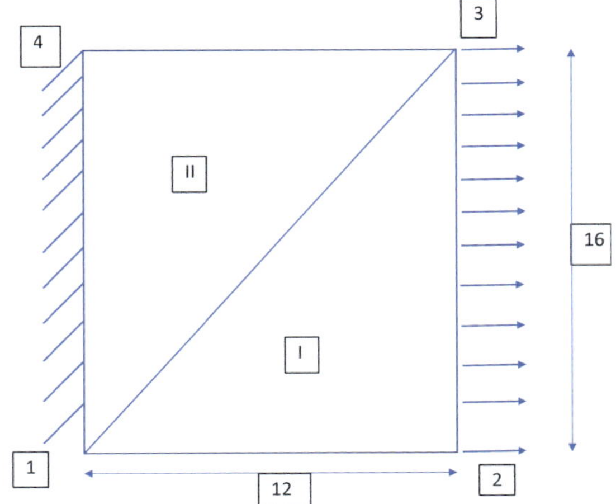

Fig. 4.20 A rectangular region with two triangular elements

After all these terms are **assembled** the element equation becomes:

$$[\bar{u}]^T[K][u] = [\bar{u}]^T[f^1] + [\bar{u}]^T[f^2] \tag{4.114}$$

Example 4.6 This example problem is a plate with length of a = 12.0″ and height of b = 16.0″ The thickness is 0.036″. A distributed load of 10 lb/in acting on the vertical side towards the right and the left side is fixed. The area is divided into two triangular elements. E = 30E6 lb/sq in and Poisson's Ratio = 0.25 (Fig. 4.20).

Element #1

$$x1 = y1 = y2 = 0, x2 = x3 = a, y3 = b \tag{4.115}$$

Using this information, we create the stiffness matrix from the following relationships:

$$N_i = \frac{1}{J}(\text{pi} + \text{qix} + \text{riy}) \tag{4.116}$$

where,

$$\text{p1} = (x_2 y_3 - x_3 y_2); \text{q1} = (y_2 - y_3); \text{r1} = (x_3 - x_2) \tag{4.117}$$

$$\text{p2} = (x_3 y_1 - x_1 y_3); \text{q2} = (y_3 - y_1); \text{r2} = (x_1 - x_3) \tag{4.118}$$

$$\text{p3} = (x_1 y_2 - x_2 y_1); \text{q3} = (y_1 - y_2); \text{r3} = (x_2 - x_1) \tag{4.119}$$

$$\frac{\partial N_1}{\partial x} = \frac{(y2-y3)}{J} = \frac{-b}{J}; \frac{\partial N_2}{\partial x} = \frac{(y3-y1)}{J} = \frac{b}{J}; \frac{\partial N_3}{\partial x} = \frac{(y1-y2)}{J} = 0 \qquad (4.120)$$

$$\frac{\partial N_1}{\partial y} = \frac{(x3-x2)}{J} = 0, \frac{\partial N_2}{\partial y} = \frac{(x1-x3)}{J} = \frac{-a}{J}, \frac{\partial N_3}{\partial y} = \frac{(x2-x1)}{J} = \frac{a}{J} \qquad (4.121)$$

The stress strain relationship changes accordingly

$$\begin{bmatrix} \sigma_{xx} \\ \sigma_{yy} \\ \sigma_{xy} \end{bmatrix} = \begin{bmatrix} C11 & C12 & 0 \\ C12 & C22 & 0 \\ 0 & 0 & C66 \end{bmatrix} \begin{bmatrix} \epsilon_{xx} \\ \epsilon_{yy} \\ \gamma_{xy} \end{bmatrix} \qquad (4.122)$$

For plane stress:

$$C11 = C22 = \frac{E}{(1-v^2)} \qquad (4.123)$$

$$C12 = \frac{vE}{(1-v^2)} \qquad (4.124)$$

$$C66 = G = \frac{E}{2(1+v)} \qquad (4.125)$$

```
C11 = 30E6/(1-0.25^2)
```

```
C11 =
   32000000
```

```
C22 = C11
```

```
C22 =
   32000000
```

```
C12 = 0.25*C11
```

```
C12 =
    8000000
```

```
C66 = 30E6/(2*(1+0.25))
```

```
C66 =
   12000000
```

```
D = [C11 C12 0;C12 C22 0;0 0 C66]
```

```
D = 3×3
   32000000    8000000          0
    8000000   32000000          0
          0          0   12000000
```

$$[B] = \begin{bmatrix} \frac{\partial N_i}{\partial x} & 0 & \frac{\partial N_j}{\partial x} & 0 & \frac{\partial N_k}{\partial x} & 0 \\ 0 & \frac{\partial N_i}{\partial y} & 0 & \frac{\partial N_j}{\partial y} & 0 & \frac{\partial N_k}{\partial y} \\ \frac{\partial N_i}{\partial y} & \frac{\partial N_i}{\partial x} & \frac{\partial N_j}{\partial y} & \frac{\partial N_j}{\partial x} & \frac{\partial N_k}{\partial y} & \frac{\partial N_k}{\partial x} \end{bmatrix} \tag{4.126}$$

```
 b = 160

b =
  160


 a = 120

a =
  120


 t = 0.036

t =
  0.036000000000000


 B1 = (1/(b*a))*[-b 0 b 0 0 0;0 0 0 -a 0 a;0 -b -a b a 0]

B1 = 3×6
  -0.008333333333333                      0    0.008333333333333  ···
                   0                      0                     0
                   0    -0.008333333333333   -0.006250000000000


 K1 = (transpose(B1)*(D)*B1)*(b*a/2)*t

K1 = 6×6
10⁵ ×
   7.679999999999999                      0   -7.679999999999999  ···
                   0    2.880000000000000    2.160000000000000
  -7.679999999999999    2.160000000000000    9.299999999999999
   1.440000000000000   -2.880000000000000   -3.600000000000000
                   0   -2.160000000000000   -1.620000000000000
  -1.440000000000000                      0    1.440000000000000
```

Element #2

$$x1 = y1 = x3 = 0, x2 = a, y3 = y2 = b \tag{4.127}$$

Using this information, we create the stiffness matrix from the following relationships:

$$N_i = \frac{1}{J}(\text{p}i + \text{q}i x + \text{r}i y) \tag{4.128}$$

where,

$$\text{p1} = (x_2 y_3 - x_3 y_2); \text{q1} = (y_2 - y_3); \text{r1} = (x_3 - x_2) \tag{4.129}$$

$$\text{p2} = (x_3 y_1 - x_1 y_3); \text{q2} = (y_3 - y_1); \text{r2} = (x_1 - x_3) \tag{4.130}$$

$$\text{p3} = (x_1 y_2 - x_2 y_1); \text{q3} = (y_1 - y_2); \text{r3} = (x_2 - x_1) \tag{4.131}$$

$$\frac{\partial N_1}{\partial x} = \frac{(y2 - y3)}{J} = 0, \frac{\partial N_2}{\partial x} = \frac{(y3 - y1)}{J} = \frac{b}{J}, \frac{\partial N_3}{\partial x} = (y1 - y2) = \frac{-b}{J} \tag{4.132}$$

$$\frac{\partial N_1}{\partial y} = \frac{(x3 - x2)}{J} = \frac{-a}{J}, \frac{\partial N_2}{\partial y} = \frac{(x1 - x3)}{J} = 0, \frac{\partial N_3}{\partial y} = \frac{(x2 - x1)}{J} = \frac{a}{J} \tag{4.133}$$

The stress strain relationship changes accordingly

$$\begin{bmatrix} \sigma_{xx} \\ \sigma_{yy} \\ \sigma_{xy} \end{bmatrix} = \begin{bmatrix} C11 & C12 & 0 \\ C12 & C22 & 0 \\ 0 & 0 & C66 \end{bmatrix} \begin{bmatrix} \epsilon_{xx} \\ \epsilon_{yy} \\ \gamma_{xy} \end{bmatrix} \tag{4.134}$$

For plane stress:

$$C11 = C22 = \frac{E}{(1 - v^2)} \tag{4.135}$$

$$C12 = \frac{vE}{(1 - v^2)} \tag{4.136}$$

$$C66 = G = \frac{E}{2(1 + v)} \tag{4.137}$$

```
  C11 = 30E6/(1-0.25^2)

C11 =
  32000000


  C22 = C11

C22 =
  32000000


  C12 = 0.25*C11

C12 =
   8000000


  C66 = 30E6/(2*(1+0.25))

C66 =
  12000000


  D = [C11 C12 0;C12 C22 0;0 0 C66]

D = 3×3
    32000000       8000000             0
     8000000      32000000             0
           0             0      12000000

  b = 160

b =
  160


  a = 120

a =
  120


  t = 0.036

t =
  0.036000000000000


  B2 = (1/(b*a))*[0 0 b 0 -b 0;0 -a 0 0 0 a;-a 0 0 b a -b]

B2 = 3×6
                  0                   0    0.008333333333333 ⋯
                  0   -0.006250000000000                    0
 -0.006250000000000                   0                    0

  K2 = (transpose(B2)*D*B2)*(b*a/2)*t

K2 = 6×6
10⁵ ×
     1.620000000000000                   0                    0 ⋯
                     0   4.319999999999999   -1.440000000000000
                     0  -1.440000000000000    7.679999999999999
    -2.160000000000000                   0                    0
    -1.620000000000000   1.440000000000000   -7.679999999999999
     2.160000000000000  -4.319999999999999    1.440000000000000
```

$$[B] = \begin{bmatrix} \dfrac{\partial N_i}{\partial x} & 0 & \dfrac{\partial N_j}{\partial x} & 0 & \dfrac{\partial N_k}{\partial x} & 0 \\[2mm] 0 & \dfrac{\partial N_i}{\partial y} & 0 & \dfrac{\partial N_j}{\partial y} & 0 & \dfrac{\partial N_k}{\partial y} \\[2mm] \dfrac{\partial N_i}{\partial y} & \dfrac{\partial N_i}{\partial x} & \dfrac{\partial N_j}{\partial y} & \dfrac{\partial N_j}{\partial x} & \dfrac{\partial N_k}{\partial y} & \dfrac{\partial N_k}{\partial x} \end{bmatrix} \tag{4.138}$$

```
Kt = [K1(3,3) K1(3,4) K1(3,5) K1(3,6);K1(4,3) K1(4,4) K1(4,5)
K1(4,6);K1(5,3),K1(5,4),K1(5,5)+K2(3,3),K1(5,6)+K2(3,4);K1(6,3)
K1(6,4),K1(6,5)+K2(4,3), K1(6,6)+K2(4,4)]
```

```
Kt = 4×4
10⁵ ×
     9.299999999999999   -3.600000000000000   -1.620000000000000 ...
    -3.600000000000000    7.200000000000000    2.160000000000000
    -1.620000000000000    2.160000000000000    9.299999999999999
     1.440000000000000   -4.319999999999999                    0
```

```
F = [10*160./2; 0; 10*160./2; 0]
```

```
F = 4×1
   800
     0
   800
     0
```

```
U = Kt\F
```

```
U = 4×1
    0.001129111438390
    0.000196367206676
    0.001011291114384
   -0.000108001963672
```

```
%format long
U
```

```
U = 4×1
    0.001129111438390
    0.000196367206676
    0.001011291114384
   -0.000108001963672
```

Once the entire displacement matrix is computed, the B matrix * displacement is the strain and the straining times the D matrix is the stress in each element.

So the displacement matrix of each element is given by (U1 is the displacement matrix of the first element and U2 is the displacement matrix of the second element).

$$U1 = \begin{bmatrix} 0 \\ 0 \\ 0.001129 \\ 0.000196 \\ 0.001011 \\ -0.000108 \end{bmatrix}$$

and

$$U2 = \begin{bmatrix} 0 \\ 0 \\ 0.001011 \\ -0.000108 \\ 0 \\ 0 \end{bmatrix}$$

We will now compute the stress vector in each element as St1 and St2. While the displacements are computed at the nodes, the stresses will be calculated over the element. There will be three components of stress, normal stresses in the x- and y-directions and the shear stress.

```
U1 =[0;0;U(1);U(2);U(3);U(4)]

U1 = 6×1
                        0
                        0
     0.001129111438390
     0.000196367206676
     0.001011291114384
    -0.000108001963672


U2 =[0;0;U(3);U(4);0;0]

U2 = 6×1
                        0
                        0
     0.001011291114384
    -0.000108001963672
                        0
                        0


%Stress vector in first element
St1= D*B1*U1

St1 = 3×1
10² ×
     2.858779250531829
     0.144002618229423
     0.108001963672067


%Stress vector in second element
St2= D*B2*U2

St2 = 3×1
10² ×
     2.696776305023728
     0.674194076255932
    -0.108001963672067
```

Example 4.7 Stress Concentration in Plate with Circular Hole
Perform a 2-D plane-stress elasticity analysis.

A thin rectangular plate under a uniaxial tension has a uniform stress distribution. Introducing a circular hole in the plate disturbs the uniform stress distribution near the hole, resulting in a significantly higher than average stress. Such a thin plate, subject to in-plane loading, can be analyzed as a 2-D plane-stress elasticity problem. In theory, if the plate is infinite, then the stress near the hole is three times higher than the average stress. For a rectangular plate of finite width, the stress concentration factor is a function of the ratio of hole diameter to the plate width. This example approximates the stress concentration factor using a plate of a finite width.

Create Structural Model and Include Geometry
Create a structural model for static plane-stress analysis.

```
model = createpde('structural','static-planestress');
```

The plate must be sufficiently long, so that the applied loads and boundary conditions are far from the circular hole. This condition ensures that a state of uniform tension prevails in the far field and, therefore, approximates an infinitely long plate. In this example the length of the plate is four times greater than its width. Specify the following geometric parameters of the problem.

```
radius = 20.0;
width = 50.0;
Xc = 0;
Yc = 0;
totalLength = 4*width;
```

Define the geometry description matrix (GDM) for the rectangle and circle.

```
R1 = [3 4 -totalLength  totalLength ...
         totalLength -totalLength ...
         -width -width width width]'; % 3 for rectangle, 4 for number of sides
     % followed by x coordinates of the four corners and then the y
     % coordinates for the four corners.
C1 = [1 Xc Yc radius 0 0 0 0 0 0]'; % the first four items deefine a circle (1
for circle,
% 0 0 for center and radius, the other zeros are to make the vector length
% same as R1
```

Define the combined GDM, name-space matrix, and set formula to construct decomposed geometry using decsg.

```
gdm = [R1 C1]; % combine the shapes into one matrix
ns = char('R1','C1'); % create names of the basic shapes
g = decsg(gdm,'R1 - C1',ns'); %decsg operation creates the geometry from gdm,
through
% Boolean operation R1-C1 and the name vector transpose.
```

Create the geometry and include it into the structural model.

```
geometryFromEdges(model,g);% the geometry is built
```

Plot the geometry displaying edge labels (Fig. 4.21).

```
figure
pdegplot(model,'EdgeLabel','on','Vertexlabel','on');
axis([-1.2*totalLength 1.2*totalLength -1.2*width 1.2*width])
title 'Geometry with Edge Labels';
```

Specify Model Parameters

Specify the Young's modulus and Poisson's ratio to model linear elastic material behavior. Remember to specify physical properties in consistent units.

```
structuralProperties(model,'YoungsModulus',200E3,'PoissonsRatio',0.25);
```

Restrain all rigid-body motions of the plate by specifying sufficient constraints. For static analysis, the constraints must also resist the motion induced by applied load.

Set the x-component of displacement on the left edge (edge 3) to zero to resist the applied load. Set the y-component of displacement at the bottom left corner (vertex 3) to zero to restraint the rigid body motion.

```
structuralBC(model,'Edge',3,'XDisplacement',0);%left edge is fixed
structuralBC(model,'Vertex',3,'YDisplacement',0);%left edge is fixed
```

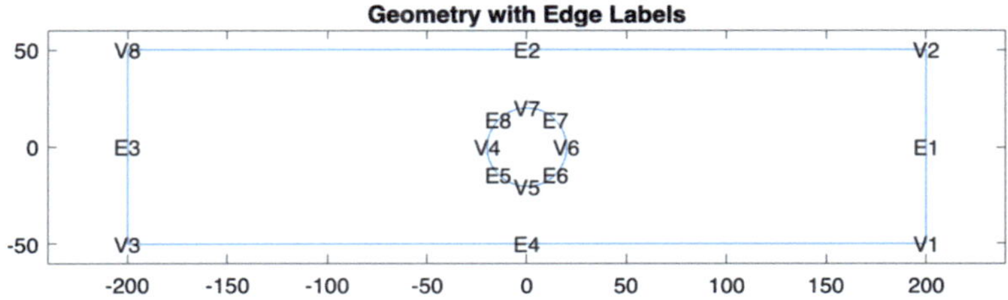

Fig. 4.21 Plate geometry

Apply the surface traction with a non-zero x-component on the right edge of the plate.

```
structuralBoundaryLoad(model,'Edge',1,'SurfaceTraction',[100;0]); %traction at
the right edge
```

Generate Mesh and Solve

To capture the gradation in solution accurately, use a fine mesh. Generate the mesh, using Hmax to control the mesh size.

```
generateMesh(model,'Hmax',radius/6); %meshing the geeomtery
```

Plot the mesh (Fig. 4.22).

```
figure
pdemesh(model)
```

Solve the plane-stress elasticity model.

```
R = solve(model);
```

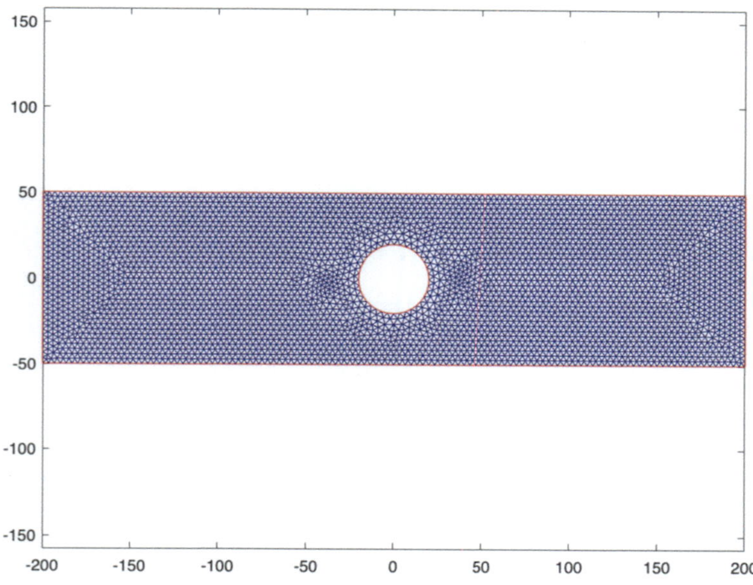

Fig. 4.22 Meshed geometry

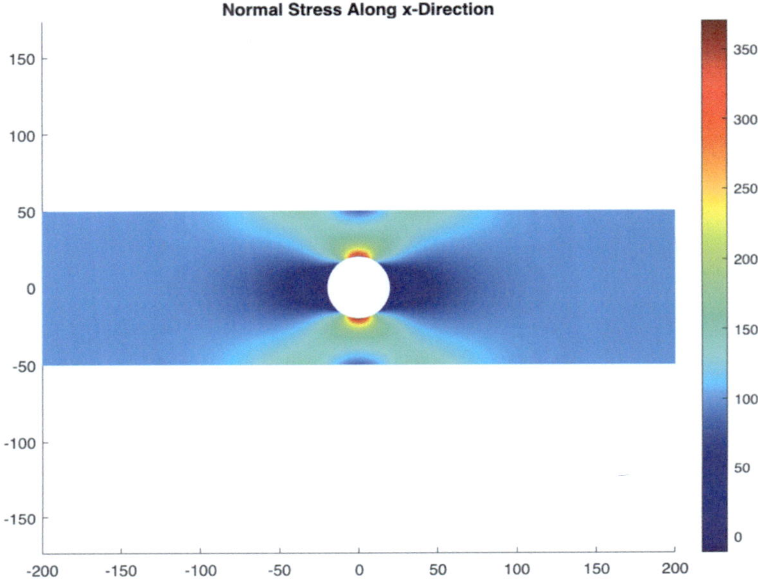

Fig. 4.23 Stress distribution

Plot Stress Contours

Plot the x-component of the normal stress distribution. The stress is equal to applied tension far away from the circular boundary. The maximum value of stress occurs near the circular boundary (Fig. 4.23).

```
R = solve(model);


figure
pdeplot(model,'XYData',R.Stress.xx,'ColorMap','jet');
axis equal
title 'Normal Stress Along x-Direction';
```

Interpolate Stress

To see the details of the stress variation near the circular boundary, first define a set of points on the boundary.

```
thetaHole = linspace(0,2*pi,200);
xr = radius*cos(thetaHole);
yr = radius*sin(thetaHole);
CircleCoordinates = [xr;yr];
```

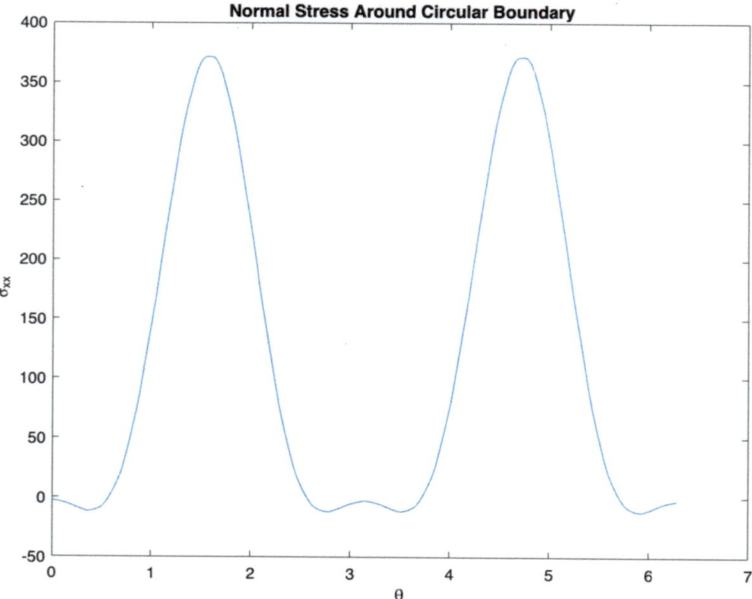

Fig. 4.24 Stress distribution around the hole boundary

Then interpolate stress values at these points by using interpolateStress. This function returns a structure array with its fields containing interpolated stress values.

```
stressHole = interpolateStress(R,CircleCoordinates);
```

Plot the normal direction stress versus angular position of the interpolation points (Fig. 4.24).

```
figure
plot(thetaHole,stressHole.sxx)
xlabel('\theta')
ylabel('\sigma_{xx}')
title 'Normal Stress Around Circular Boundary';
```

Solve the same problem using a symmetric model
The plate with a hole model has two axes of symmetry. Therefore, you can model a quarter of the geometry. The following model solves a quadrant of the full model with appropriate boundary conditions. Another difference from the previous version is that instead of a uniform traction, a traction that varies with location is used in the form of a user defined function.

Create a structural model for the static plane-stress analysis.

```
symModel = createpde('structural','static-planestress');%structural model
```

Create the geometry that represents one quadrant of the original model. You do not need to create additional edges to constrain the model properly.

```
radius = 20.0;
width = 50.0;
Xc = 0;
Yc = 0;
totalLength = 4*width;
R1 = [3 4 0 totalLength/2 totalLength/2 ...
      0 0 0 width width]';
C1 = [1 0 0 radius 0 0 0 0 0 0]';
gm = [R1 C1];
sf = 'R1-C1';
ns = char('R1','C1');
g = decsg(gm,sf,ns');
geometryFromEdges(symModel,g);
```

Plot the geometry displaying the edge labels (Fig. 4.25).

```
figure
pdegplot(symModel,'EdgeLabel','on');
axis equal
title 'Symmetric Quadrant with Edge Labels';
```

Specify structural properties of the material.

```
structuralProperties(symModel,'YoungsModulus',200E3, ...
                     'PoissonsRatio',0.25); % Structural material
properties
```

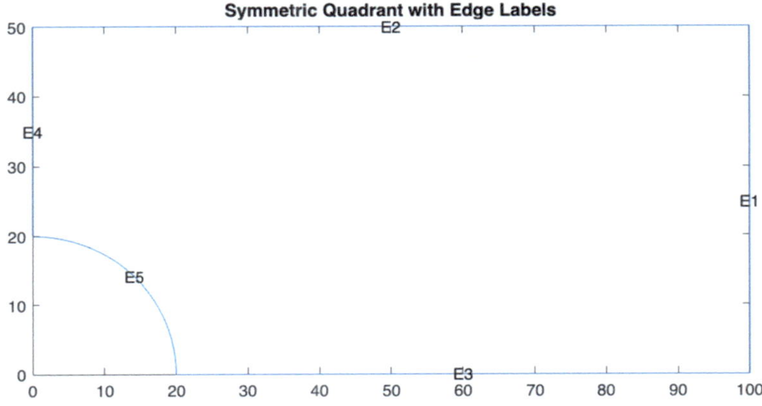

Fig. 4.25 Symmetric geometry

Apply symmetric constraints on the edges 3 and 4.

```
structuralBC(symModel,'Edge',[3 4],'Constraint','symmetric');%fixed boundary
condition
```

Apply surface traction on the edge 1.

```
structuralBoundaryLoad(symModel,'Edge',1,'SurfaceTraction',@tracFcn);%traction
load is varying and it is incorporated using a function
```

Generate mesh and solve the symmetric plane-stress model.

```
generateMesh(symModel,'Hmax',radius/6);
Rsym = solve(symModel);
```

Plot the x-component of the normal stress distribution. The results are identical to the first quadrant of the full model (Fig. 4.26).

```
figure
pdeplot(symModel,'XYData',Rsym.Stress.xx,'ColorMap','jet');
axis equal
title 'Normal Stress Along x-Direction for Symmetric Model';
```

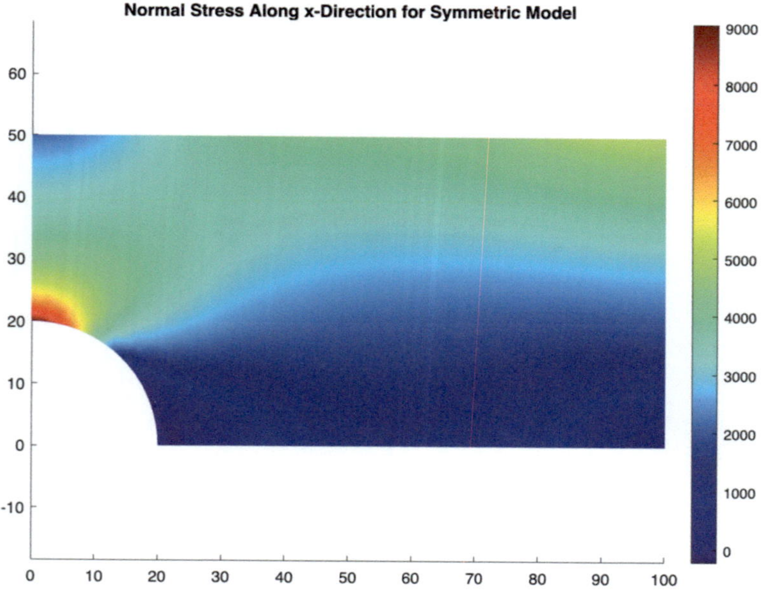

Fig. 4.26 Stress distribution on the quarter plate

```
function trac = tracFcn(location,state) %function for traction

trac = [100*location.y,0]; %the x-force on the right edge increases with the y
coordinate.

end
```

Example 4.8 Twisting a Lego. In this example we import a stl file of a Lego block, a relatively complex geometry, and apply a torque at one end of the block using two opposite forces at two corner vertices. The other end of the Lego is fixed (Fig. 4.27).

```
model = createpde('structural','static-solid');%create a static analysis model
 importGeometry(model,'Lego.stl');%import the Lego block geometry from the stl
file
 figure % the dimensions are in mm
 pdegplot(model,"FaceLabels","on","VertexLabels","on","FaceAlpha",0.5)%draw the
geometry with faces and edges shown
```

```
E = 2E3; % the units are N/mm^2
nu = 0.3;
structuralProperties(model,'YoungsModulus',E, ...
                     'PoissonsRatio',nu);
```

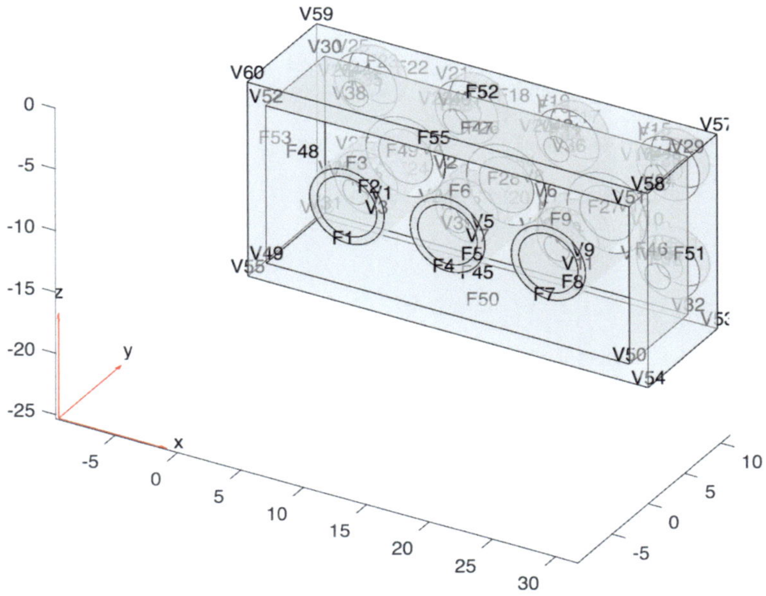

Fig. 4.27 Lego brick geometry

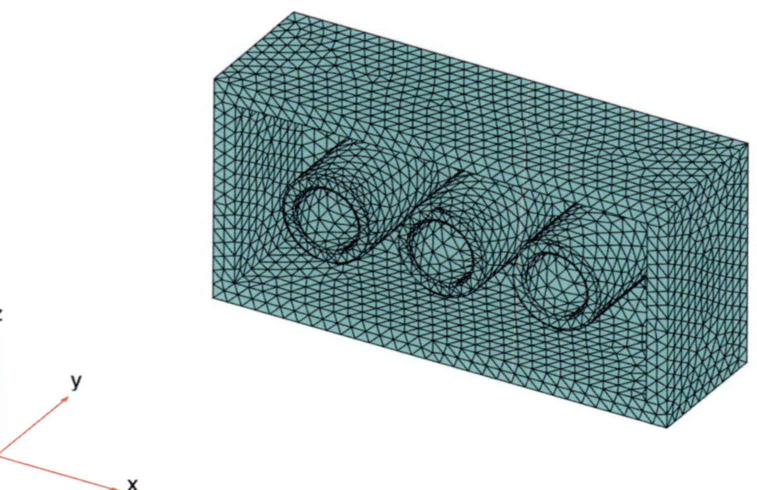

Fig. 4.28 Meshed geometry

Generate a mesh (Fig. 4.28).

```
generateMesh(model,'Hmax',1); %mesh is created

pdemesh (model); %display the model and the mesh
```

```
structuralBC(model,'Face',53,'Constraint','fixed');%left edge is fixed
structuralBoundaryLoad(model,'Vertex',58,'Force',[0;50;0]);%apply a 50 N force
on one corner
structuralBoundaryLoad(model,'Vertex',53,'Force',[0;-50;0]);% apply a -50N
force on another corner.
% together the two forces are forming a torque
```

```
R = solve(model); %solve the model and save the results in R
figure
pdeplot3D(model,"ColorMapData",R.Stress.yz,"FaceAlpha",.3); %view the yz or
shear stress
view([60.3 4.9])
title 'Shear Stress in yz-Plane';
```

The shear stress and the displacement plots are shown in Figs. 4.29 and 4.30 respectively.

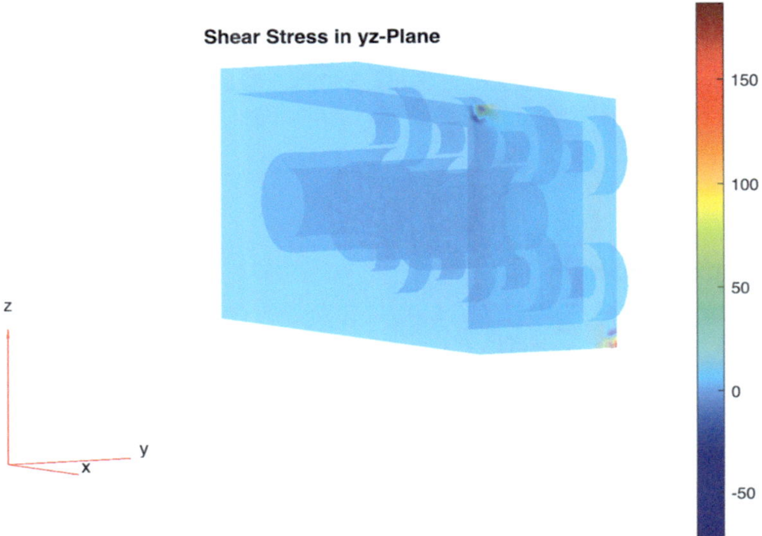

Fig. 4.29 Shear stress in the plane

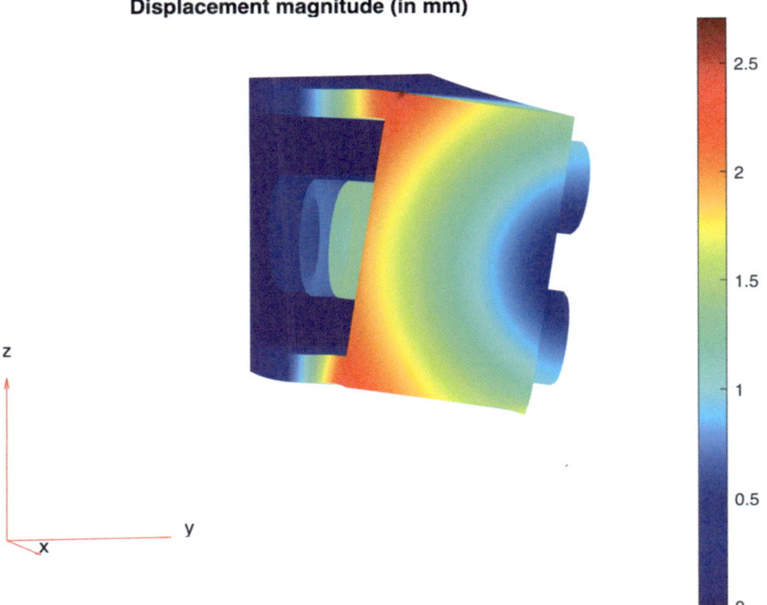

Fig. 4.30 Deformed geometry

```
pdeplot3D(model,"ColorMapData",R.Displacement.Magnitude,"Deformation",R.Displac
ement, "DeformationScaleFactor",1)
axis equal
title 'Displacement magnitude (in mm) ';

view([78.6 5.1])
```

4.3   Dynamic Loading

Example 4.9 This is an example of dynamic loading. In this example we will tackle a number of analyses. The geometry is a 3-D hollow vertical cantilever, approximating a tall building. The first part of the analysis is a dynamic model to find the natural frequencies of vibration and modes. The second part of the example looks at the effect of an initial displacement on the dynamic behavior of the beam over time, i.e. how does it oscillate after it has been initially disturbed. For this dynamic analysis we need the modal analyses results from the first part. Figures 4.31 and 4.32 shows the beam geometry and the meshed geometry, respectively.

Fig. 4.31 Beam geometry

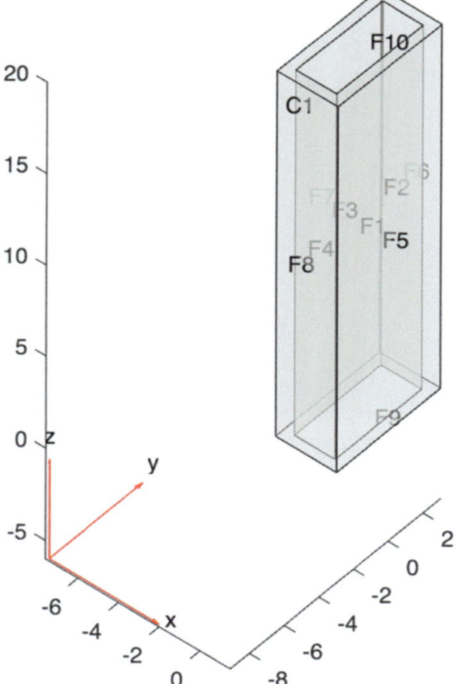

Fig. 4.32 Meshed beam

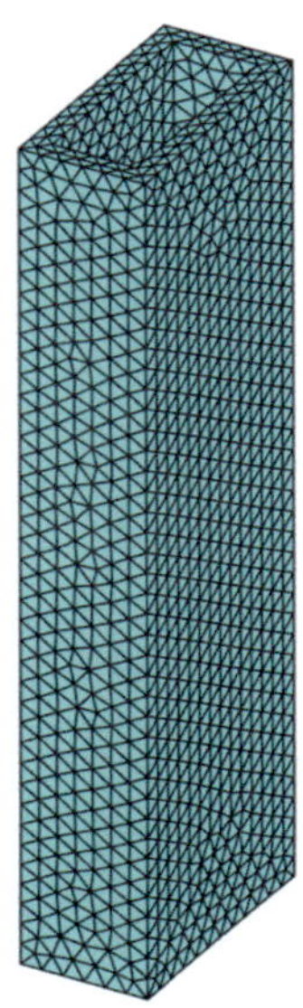

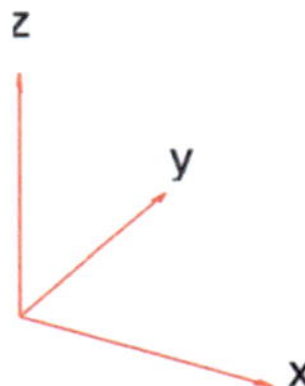


```matlab
cantileverbeam = createpde('structural','modal-solid'); %model definition for
3-D modal analysis
 gm = multicuboid([2 3],[5 6],20,'Void',[true,false]);% the geometry is created
using the
 % multicuboid command that builds a hollow box shape
 % We are using SI units, so the height of the cantilever is 20 m.


 cantileverbeam.Geometry = gm; %the geometry is assigned to the model
 pdegplot(cantileverbeam,'CellLabels','on','Facelabels','on','FaceAlpha',0.5);
%the basic geometry is drawn

 xlim([-7.6 1.5])
 ylim([-9.1 3.0])
 zlim([-6.0 20.0])
 view([40.3 45.2])


 generateMesh(cantileverbeam,'Hmax',0.5,'GeometricOrder','linear'); % generate
linear triangular elements
 figure
 pdeplot3D(cantileverbeam,"Mesh","on") %plot the mesh
 axis equal
```

```matlab
E = 210E9;% Modulus of Elasticity
nu = 0.3;% Poisson's Ratio
rho = 8000;% Mass Density
structuralProperties(cantileverbeam,'YoungsModulus',E, ...
                    'PoissonsRatio',nu, ...
                    'MassDensity',rho);% Assign properties to the model
structuralBC(cantileverbeam,'Face',9,'Constraint','fixed'); % The base of the
beam is cantilevered.
```

Solve the model for the natural frequencies. Specify the lower frequency limit below zero so that all modes with frequencies near zero appear in the solution.

```matlab
RF = solve(cantileverbeam,'FrequencyRange',[-1,280]*2*pi);% the calculation
%is limited by the 280 (chosen arbitrarily). Thee results are assigned to
%the variable RF
```

By default, the solver returns circular frequencies.

```matlab
modeID = 1:numel(RF.NaturalFrequencies);
```

Express the resulting frequencies in Hz by dividing them by 2π. Display the frequencies in a table. Also, the first eight natural frequencies and modes are shown in Fig. 4.33.

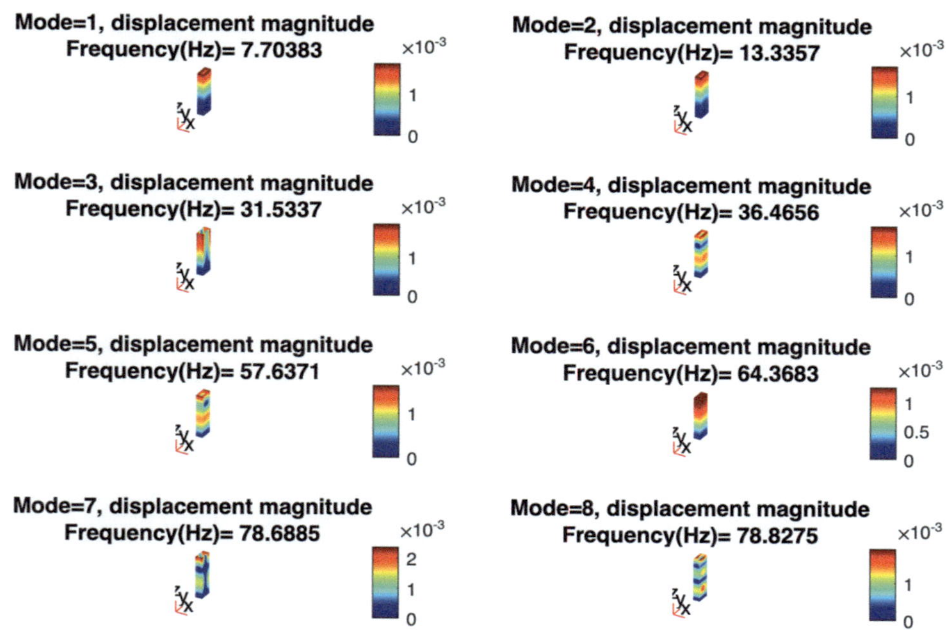

Fig. 4.33 First eight modes

```
freqHz = RF.NaturalFrequencies/2/pi; %frequencies in Hz
tmodalResults = table(modeID.',freqHz); % preping for table display
tmodalResults.Properties.VariableNames = {'Mode','Frequency'};
disp(tmodalResults);
```

Mode	Frequency
1	7.7038
2	13.336
3	31.534
4	36.466
5	57.637
6	64.368
7	78.689
8	78.827
9	98.654
10	109.04
11	116
12	117.74
13	120.17
14	124.1
15	134.02
16	134.58
17	158.01
18	161.95
19	168.02
20	179.46
21	184.59
22	186.04
23	189.04
24	190.65
25	199.16
26	206.57
27	211.82
28	226.89
29	227.68
30	229.81
31	231.35
32	257.88
33	258.36
34	266.32
35	271.09
36	272.71

```
ans = 3078×1
        0
```

```
        0
   0.0017
   0.0017
        0
   0.0017
        0
   0.0017
        0
        0
    ⋮
```

```
% we draw some of the mode shapes (specifically the first 8) using color
% contours
h = figure;
h.Position = [100,100,900,600];
numToPrint = 8;
for i = 1:numToPrint
    subplot(4,2,i);
    pdeplot3D(cantileverbeam,'ColorMapData',RF.ModeShapes.Magnitude(:,i));
    axis equal
    title(sprintf(['Mode=%d, displacement magnitude\n', ...
    'Frequency(Hz)= %g'], ...
    i,freqHz(i)));
end
```

Initial Displacement from Static Solution

In this part of the analysis we look at the beam response when it is deformed by applying
an external load on one of its faces and immediately released at time $t = 0$. Find the initial
condition for the transient analysis by using the static solution of the beam with a load on
a face. Figures 4.34 and 4.35 show the beam geometry and the mesh, respectively.

```
modelS = createpde('structural','static-solid');% static model is defined
% with the same cantilever model
gdm = multicuboid([2 3],[5 6],20,'Void',[true,false]);
```

```
gdm =
  DiscreteGeometry with properties:

    NumCells: 1
    NumFaces: 10
    NumEdges: 24
    NumVertices: 16
       Vertices: [16×3 double]
```

Fig. 4.34 Beam geometry

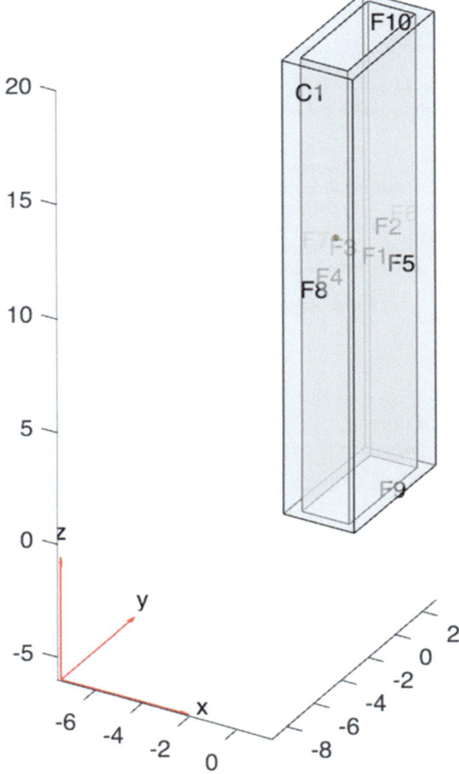

Fig. 4.35 Meshed geometry

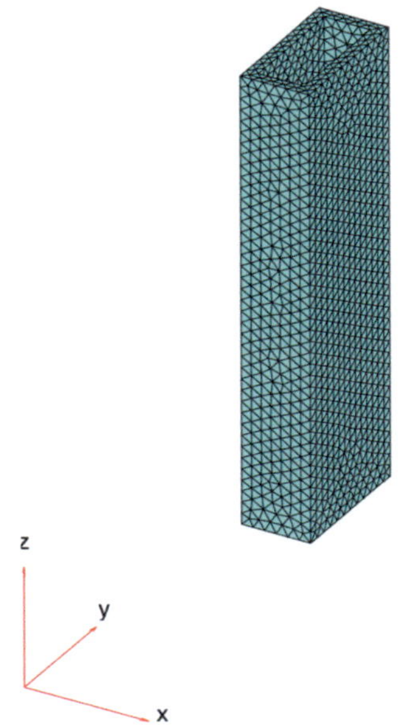

```
modelS.Geometry=gdm; %assign the geeometry
```

modelS =
 StructuralModel with properties:

         AnalysisType: "static-solid"
             Geometry: [1×1 DiscreteGeometry]
     MaterialProperties: []
             BodyLoads: []
     BoundaryConditions: []
    ReferenceTemperature: []
  SuperelementInterfaces: []
                  Mesh: []
         SolverOptions: [1×1 pde.PDESolverOptions]

```
figure
pdegplot(modelS,'CellLabels','on','Facelabels','on','FaceAlpha',0.5); %draw the
schematic
```

```
msh= generateMesh(modelS,'Hmax',0.5,'GeometricOrder','linear'); % generate
linear triangular elements
figure
pdeplot3D(modelS,"Mesh","on") %plot the mesh
axis equal
```

```
structuralProperties(modelS,'YoungsModulus',E, ...
                           'PoissonsRatio',nu, ...
                           'MassDensity',rho); % assign the material properties
that were defined before
structuralBC(modelS,'Face',9,'Constraint','fixed'); % thee base is fixed

structuralBoundaryLoad(modelS,'Face',5,'SurfaceTraction',[-10;0;0]); % a
surface traction is applied to
% a face towards the negative X direction.
Rstatic = solve(modelS); % solve for static soution
pdeplot3D(modelS,"Deformation",Rstatic.Displacement)% plot the static
displacement
```

Figures 4.36 and 4.37 show the static displacement and VonMises stress distribution, respectively for this static displacement analysis.

```
pdeplot3D(modelS,"ColorMapData",Rstatic.VonMisesStress) % display the VonMises
stress
```

Transient Analysis

Perform the transient analysis of the cantilever beam with and without damping. Use the modal superposition method to speed up computations.

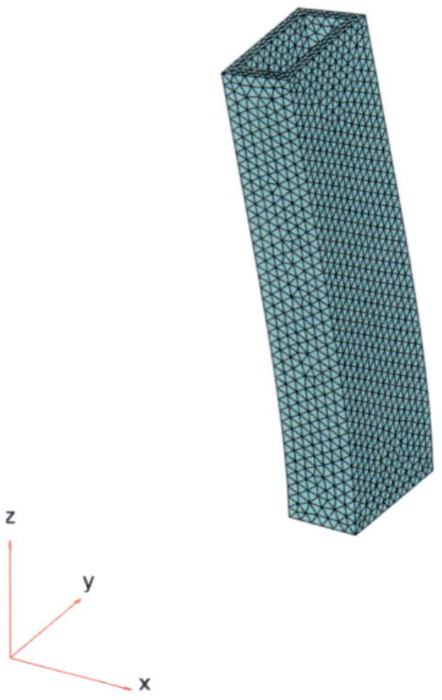

Fig. 4.36 Static displacement of beam

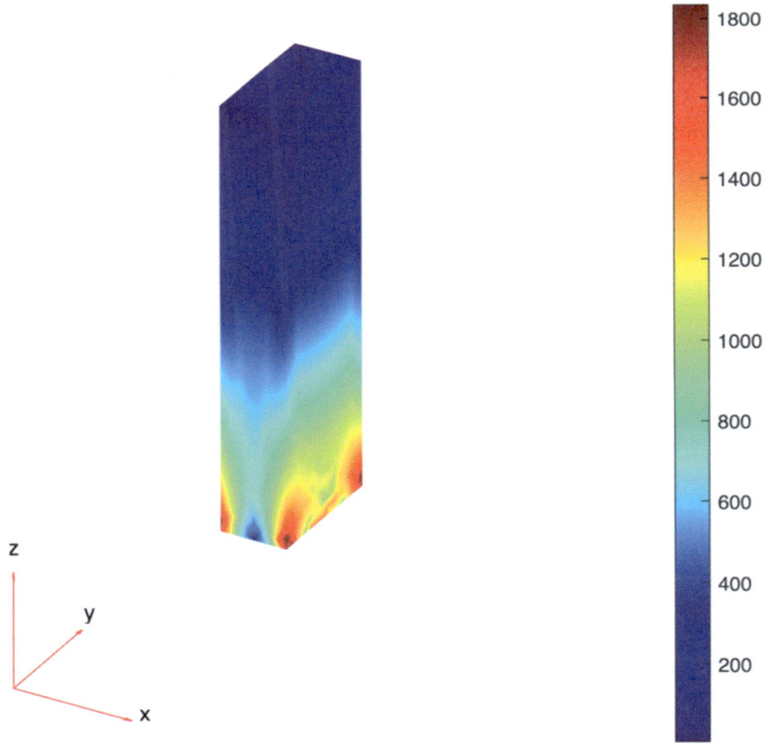

Fig. 4.37 VonMises stress in static analysis

Create a transient plane-stress model.

```matlab
modelT = createpde('structural','transient-solid'); %create a transient model
```

Use the same geometry and mesh that you used for the modal analysis (Figs. 4.38 and 4.39).

```matlab
modelT.Geometry=gdm;% assign geometry
```

```
modelT =
  StructuralModel with properties:

            AnalysisType: "transient-solid"
                Geometry: [1×1 DiscreteGeometry]
      MaterialProperties: []
               BodyLoads: []
       BoundaryConditions: []
            DampingModels: []
         InitialConditions: []
   SuperelementInterfaces: []
                     Mesh: []
            SolverOptions: [1×1 pde.PDESolverOptions]
```

```matlab
figure
pdegplot(modelT,'CellLabels','on','Facelabels','on','FaceAlpha',0.5) %draw
gometry
```

```matlab
modelT.Mesh = msh;
%%figure
pdeplot3D(modelT,"Mesh","on") %plot the mesh
axis equal
```

```matlab
structuralProperties(modelT,'YoungsModulus',E, ...
                            'PoissonsRatio',nu, ...
                            'MassDensity',rho); % assign material properties
structuralBC(modelT,'Face',9,'Constraint','fixed'); % fix the base surface
structuralIC(modelT,Rstatic);% apply the static solution from previous run as
initial conditions

tlist=0:0.001:5;% the time for the transient solution is defined
resT = solve(modelT,tlist,"ModalResults",RF);%solve for the transient results
pdemesh(modelT,'NodeLabels','on'); %draw the model to pick ann aappropriate
%node number to draw displacement
xlim([-2,0])
ylim([0,3])
zlim([18,20])
```

Fig. 4.38 Beam geometry

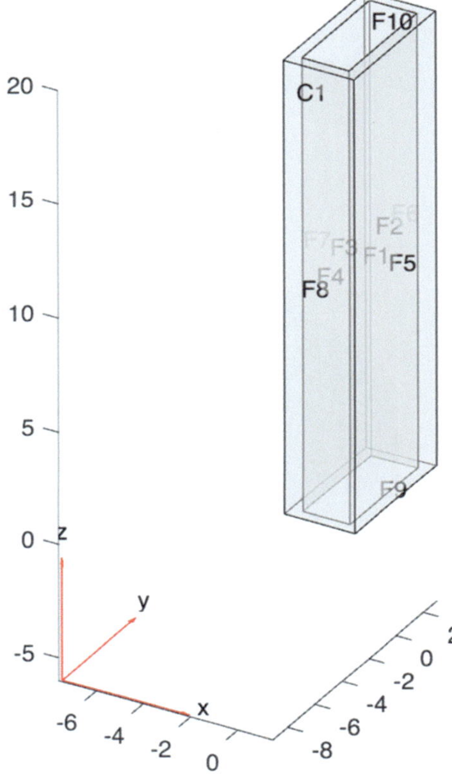

Fig. 4.39 Meshed beam

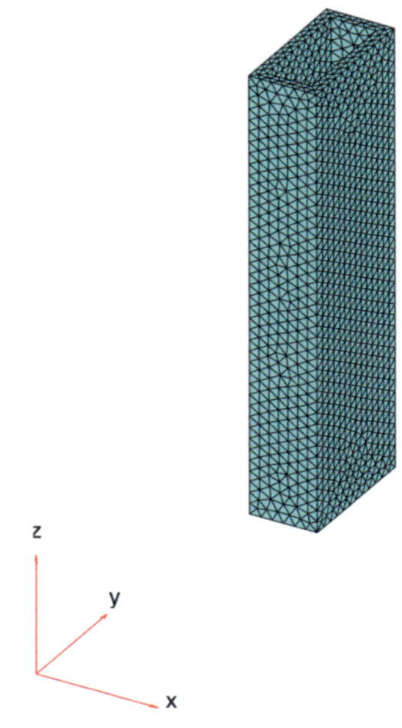

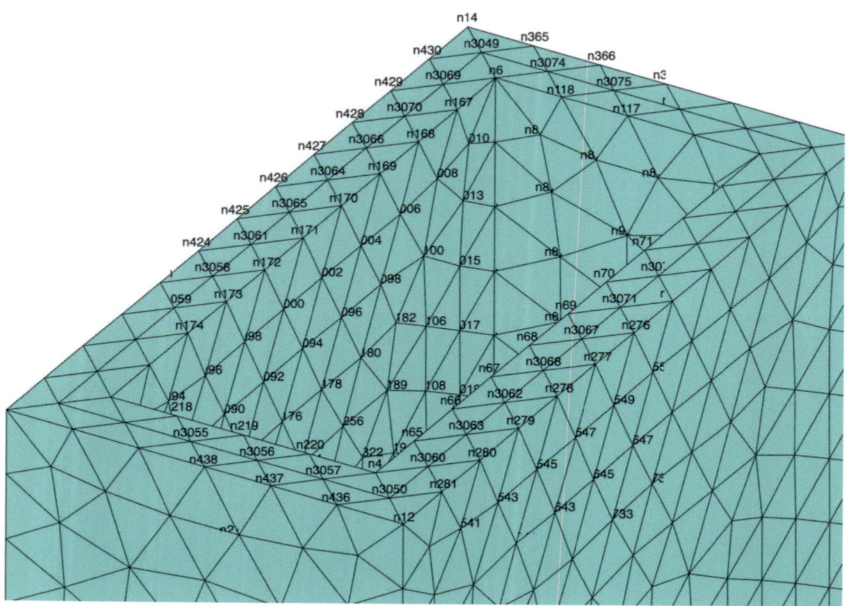

Fig. 4.40 Beam tip with node numbers

A close-up of the beam tip is drawn to pick the node numbers that will be tracked for dynamic behavior (Fig. 4.40).

```
plot(resT.SolutionTimes,resT.Displacement.ux(426,:))
grid on
title('Undamped Solution')
xlabel('Time')
ylabel('Tip of beam displacement')
```

Undamped and damped dynamic behavior of the beam are shown in Figs. 4.41 and 4.42, respectively.

```
zeta = 0.03;
%omega = 2*pi*freqNumerical;
structuralDamping(modelT,'Zeta',zeta);
resT = solve(modelT,tlist,'ModalResults',RF);
%Interpolate the displacement at the tip of the beam.

%intrpUt = interpolateDisplacement(resT,[5;0.05]);
%The y-displacement at the tip is a sinusoidal function of time with amplitude
exponentially decreasing with time.

figure
plot(resT.SolutionTimes,resT.Displacement.ux(426,:))
grid on
title('Damped Solution')
xlabel('Time')
ylabel('Tip of beam displacement')
```

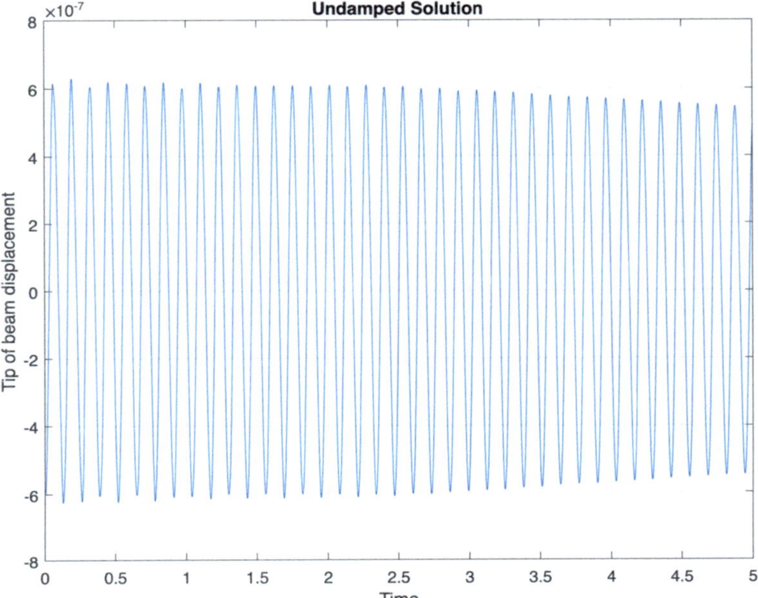

Fig. 4.41 Undamped oscillation of beam

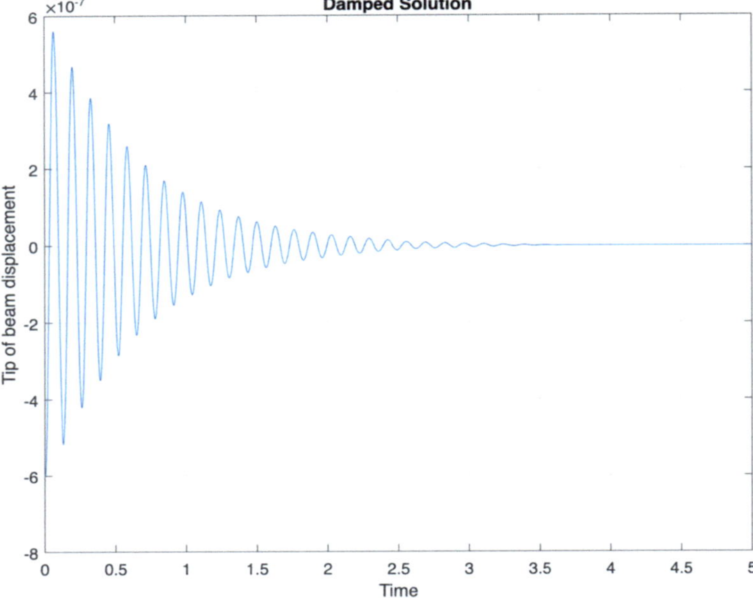

Fig. 4.42 Damped oscillation of beam

4.4 Summary

In this chapter we have covered a lot of ground. We considered a variety of Stress-analysis problems with continuum finite elements in one, two and three dimensions. With brief introduction on the formulation of beam problems and plane elasticity problems we have looked at several examples where stress analysis problems are solved for static and dynamic loading conditions.

Special Problems

5

In the previous two chapters we explored FEA for continuum problems with applications in two broad areas of stress analysis and heat transfer. In this chapter we look at three examples in detail that explore some of the advanced capabilities and applications of the PDE toolbox. The first two examples deal with thermal stress analysis for two typical manufacturing problems, quenching and welding. In both these cases a thermal analysis is performed first to generate a temperature profile history for the part. The thermal history is then used as an input to the structural analysis of the same problem to track the development of thermal and residual stresses. The third example is on electromagnetics and calculation of magnetic field and flux density for a typical application.

5.1 Thermal Stresses in Quenching

Background Information

Quenching is a material processing technique that falls under the broad category of heat treatment. When metal (e.g., steel) is heated to a high temperature and cooled very quickly by dipping it in water or oil the hot metal cools very quickly and the phase transformation of the steel, particularly on its surface, leads to the formation of a very hard phase, such as martensite. This technique was known to men in the ancient times and in the Middle Ages when weapons such as swords and spears where hardened in this manner.

This is a unique multi-physics problem since the fast-cooling rate in conjunction with the expansion and contraction that happens within the component due to the differential temperature distribution within the sample leads to residual stress formation that could be quite significant. Typically, the thermal and stress calculation in these types of problems is done using a two-step process, a transient thermal analysis that maps a history of the spatial temperature profile of the sample, followed by a stress analysis of the same model

© The Author(s), under exclusive license to Springer Nature Switzerland AG 2023   193
S. Das, *An Introduction to Finite Element Analysis Using Matlab Tools*,
Synthesis Lectures on Mechanical Engineering,
https://doi.org/10.1007/978-3-031-17540-4_5

using the temperature profile history as an input. This is a valid approach rather than a more complex coupled analysis approach because while the thermal behavior affects the structural response, the influence of the structural analysis on the thermal field can be neglected.

Here we will demonstrate the process of analysis for a representative problem. The geometry chosen is a long hollow tube. Only its cross section is modeled as a 2-D model. We assume that the boundary condition is not varying along the length of the tube. Also, since the tube is much longer in comparison to its cross-sectional dimension it is appropriate to analyze the cross-section as a plane strain section.

Create Geometry of the cross-section of a hollow tube

Create the model for Thermal analysis.

```
modelT = createpde('thermal','transient');
```

Specify the dimensions and geometric structure.

```
% Radii
rad1 = 0.2; % Dimension in m
rad2 = 0.5; % Dimension in m
% Center location
Xc = 0;
Yc = 0;
```

Define the geometry description matrix (GDM) for the rectangle and circle.

```
C1 = [1 Xc Yc rad1]'; % The two circular geometries are defined
C2 = [1 Xc Yc rad2]';
gdm = [C1 C2]; % combine the shapes into one matrix
ns = char('C1','C2'); % create names of the basic shapes
g = decsg(gdm,'C2 - C1',ns'); %decsg operation creates the geometry from gdm,
through
% Boolean operation C1-C2 and the name vector transpose.
```

Create the geometry and include it into the thermal model.

```
geometryFromEdges(modelT,g);
```

Plot the geometry with the edge and face labels (Fig. 5.1).

Fig. 5.1 Geometry of the tube

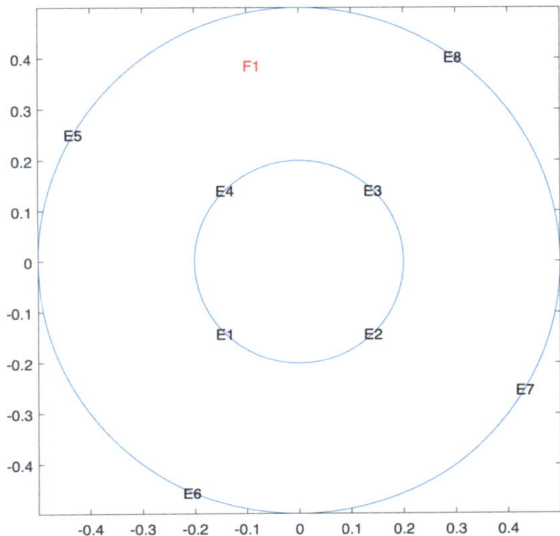

Fig. 5.2 Meshed geometry of the tube

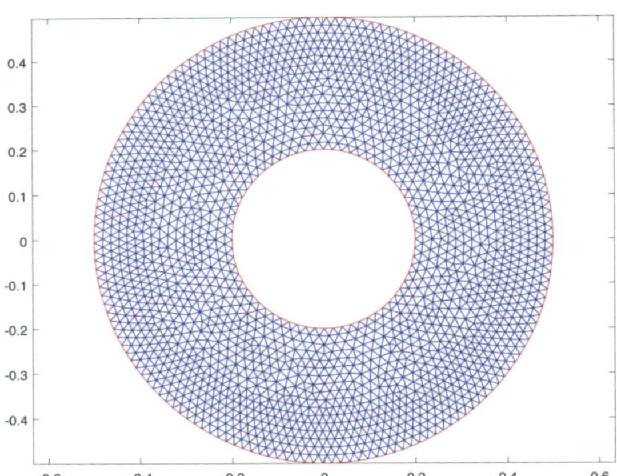

```
figure
pdegplot(modelT,'EdgeLabels','on','FaceLabels','on')
```

Generate a mesh. Use linear geometric order for elements. An approximate maximum size of the element is chosen here (Fig. 5.2).

```
maxsize=0.02

 maxsize = 0.0200
```

```
generateMesh(modelT,'Hmax',maxsize,'GeometricOrder','linear');
figure
pdemesh(modelT);
```

Specify the thermal material properties of the disc. The three properties needed in the analysis are Specific heat, thermal conductivity and mass density.

```
alphad = 1.44E-5; % Diffusivity of disc
Kd = 51;% thermal conductivity, in W/mK
rhod = 7100;% density, in Kg/m^3
cpd = Kd/rhod/alphad; % Specific Heat, J/kgK
thermalProperties(modelT,'ThermalConductivity',Kd, ...
                        'MassDensity',rhod, ...
                        'SpecificHeat',cpd);
```

Specify the convective boundary condition to account for the surface heat loss. During quenching the convective boundary conditions can be implemented using a convective heat transfer coefficient, h. In typical heat transfer problems, a constant value of h is used. There is established theory and empirical relationships available in Heat Transfer literature that provides a good estimate of what this h would be in given situations. However, the quenching problem is a little bit of an extreme heat transfer problem. When a very hot piece of metal is dipped in a cold fluid the surface heat transfer coefficient does not remain a constant. Due to boiling and evaporation that happens at hot surface a thin layer of vaporized liquid is formed at the surface shortly after the dipping. So essentially the heat transfer coefficient is a time dependent quantity. It starts out being a high value and then, when vaporization starts to happen it drops to a smaller value since the heat transfer from the surface happens through a layer of vapor rather than through liquid. To explore that we have created a time dependent function for the surface heat transfer coefficient.

For the definition of the hFcn function, see Heat Flux Function help section in MATLAB.

```
thermalBC(modelT,'Edge',[1:8],"ConvectionCoefficient",@hFcn,"AmbientTemperature
",50);
```

Set the initial temperature.

```
thermalIC(modelT,1000);
```

Solve the model for the times used in [1].

```
tlist = [0,1,10,20,50,100,200,300,500,1000,1500];
Rt = solve(modelT,tlist);
```

Plot the temperature variation with time at three key radial locations (Fig. 5.3).

```
iTRd = interpolateTemperature(Rt,[0.2;0],1:numel(Rt.SolutionTimes));
iTRp = interpolateTemperature(Rt,[0.3;0],1:numel(Rt.SolutionTimes));
iTRq = interpolateTemperature(Rt,[0.45;0],1:numel(Rt.SolutionTimes));

figure
plot(tlist,iTRd)
hold on
plot(tlist,iTRp)
plot(tlist,iTRq)
title('Temperature Variation with Time at Key Radial Locations')
legend('R_d','R_p','R_q')
xlabel 't, s'
ylabel 'T,^{\circ}C'
```

Fig. 5.3 Temperature versus time at three radial locations

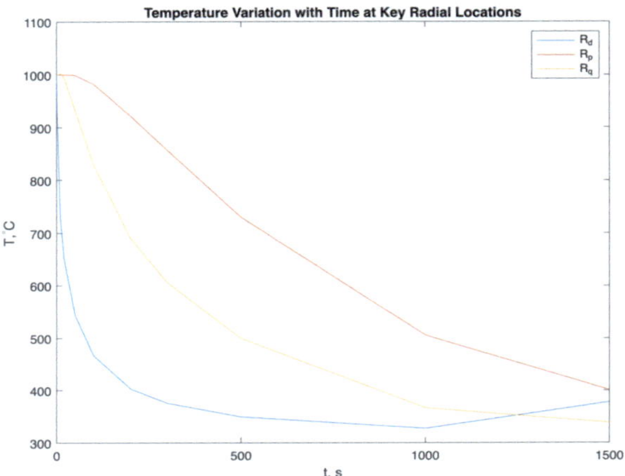

Structural Analysis: Compute Thermal Stress
Create a plane strain static structural analysis model.

```
model = createpde('structural','static-planestrain');
```

Assign the geometry and mesh used for the thermal model.

```
model.Geometry = modelT.Geometry;
model.Mesh = modelT.Mesh;
```

Specify the structural properties of the disc.

```
structuralProperties(model,'YoungsModulus',99.97E9, ...
                           'PoissonsRatio',0.29, ...
                           'CTE',1.08E-5);
```

Constrain the model to prevent rigid motion.

```
%structuralBC(model,'Edge',[1,2,3,4],"XDisplacement",0,"YDisplacement",0);
%structuralBC(model,'Edge',[1,2,3,4],'Constraint','multipoint')
structuralBC(model,'Edge',[1,2,3,4],"Constraint","roller")

ans =
  StructuralBC with properties:

          RegionType: 'Edge'
            RegionID: [1 2 3 4]
          Vectorized: 'off'

   Boundary Constraints and Enforced Displacements
           Displacement: []
          XDisplacement: []
          YDisplacement: []
          ZDisplacement: []
             Constraint: "roller"
                 Radius: []
              Reference: []
                  Label: []

   Boundary Loads
                  Force: []
        SurfaceTraction: []
               Pressure: []
   TranslationalStiffness: []
                  Label: []
```

```
%structuralBC(model,'Reference',[0,0],'Radius',20)
```

Specify the reference temperature that corresponds to the state of zero thermal stress of
the model. We use the maximum temperature as the zero-stress temperature since the

whole sample has been raised to this temperature uniformly so that the stress buildup is zero. Now when things are cooled in a non-uniform fashion the stress build-up happens.

```matlab
model.ReferenceTemperature = 1000;
```

Specify the thermal load by using the transient thermal results Rt. The solution times are the same as in the thermal model analysis. For each solution time, solve the corresponding static structural analysis problem and plot the temperature distribution, radial stress, hoop stress, and von Mises stress. For the definition of the plotResults function, see Plot Results Function. The results are comparable to Fig. 5 from (Figs. 5.4 and 5.5).

```matlab
for n = 2:numel(Rt.SolutionTimes)
structuralBodyLoad(model,'Temperature',Rt,'TimeStep',n);
R = solve(model);
plotResults(model,R,modelT,Rt,n);
end


% Tracking the evolution of VonMises Stress over time.
figure
for n = 2:numel(Rt.SolutionTimes)
%Interpolate Stress
% To see the details of the stress variation near the circular boundary, first
define a set of points on the boundary.
structuralBodyLoad(model,'Temperature',Rt,'TimeStep',n);
R = solve(model);
radialdistance = linspace(0.2,0.5,100);
xr = radialdistance;
yr = 0*radialdistance;
Coordinates = [xr;yr];
%Then interpolate stress values at these points by using interpolateStress.
This function returns a structure array with its fields containing interpolated
stress values.

stressradius = interpolateVonMisesStress(R,Coordinates);

%Plot the normal direction stress versus angular position of the interpolation
points.


plot(radialdistance,stressradius,'LineWidth',2)
hold on
xlabel('Radial distance')
ylabel('Von Mises Stress')
title 'Evolution of Von Mises Stress Along a Radial Line';
end
```

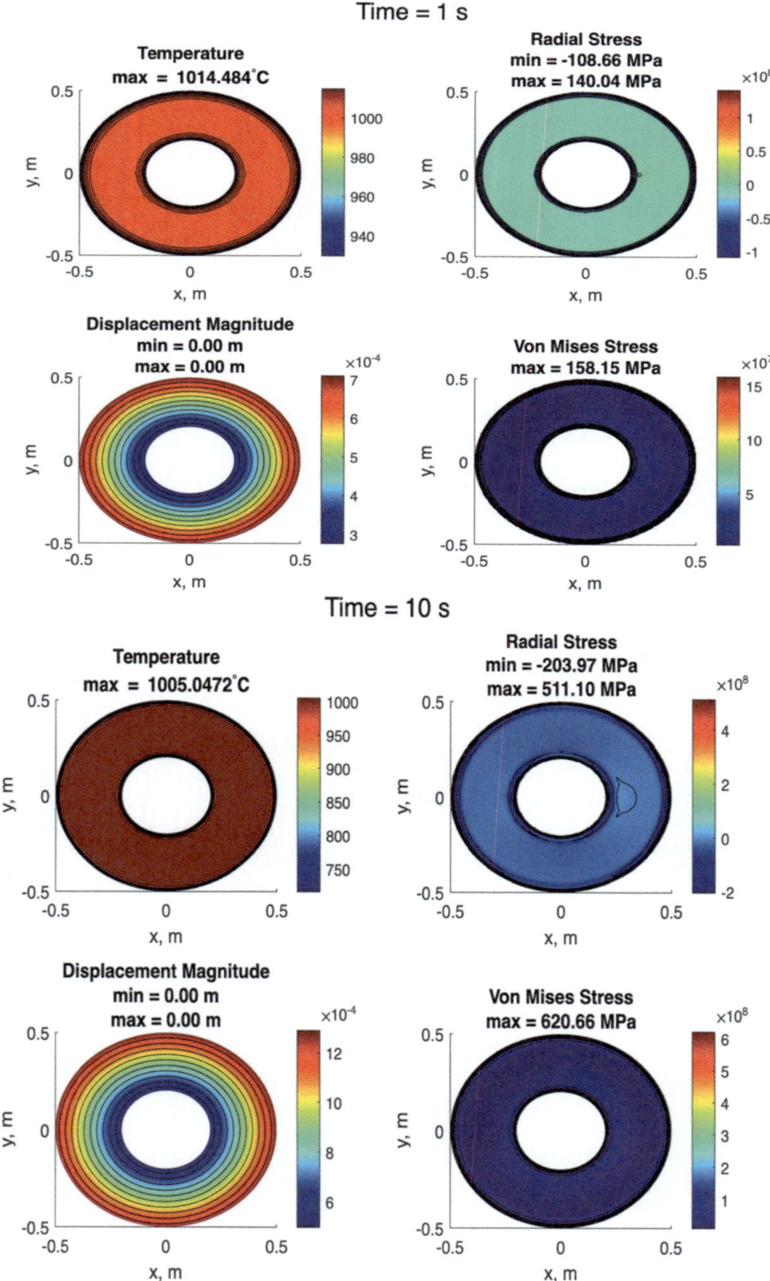

Fig. 5.4 Key solution contours at different times

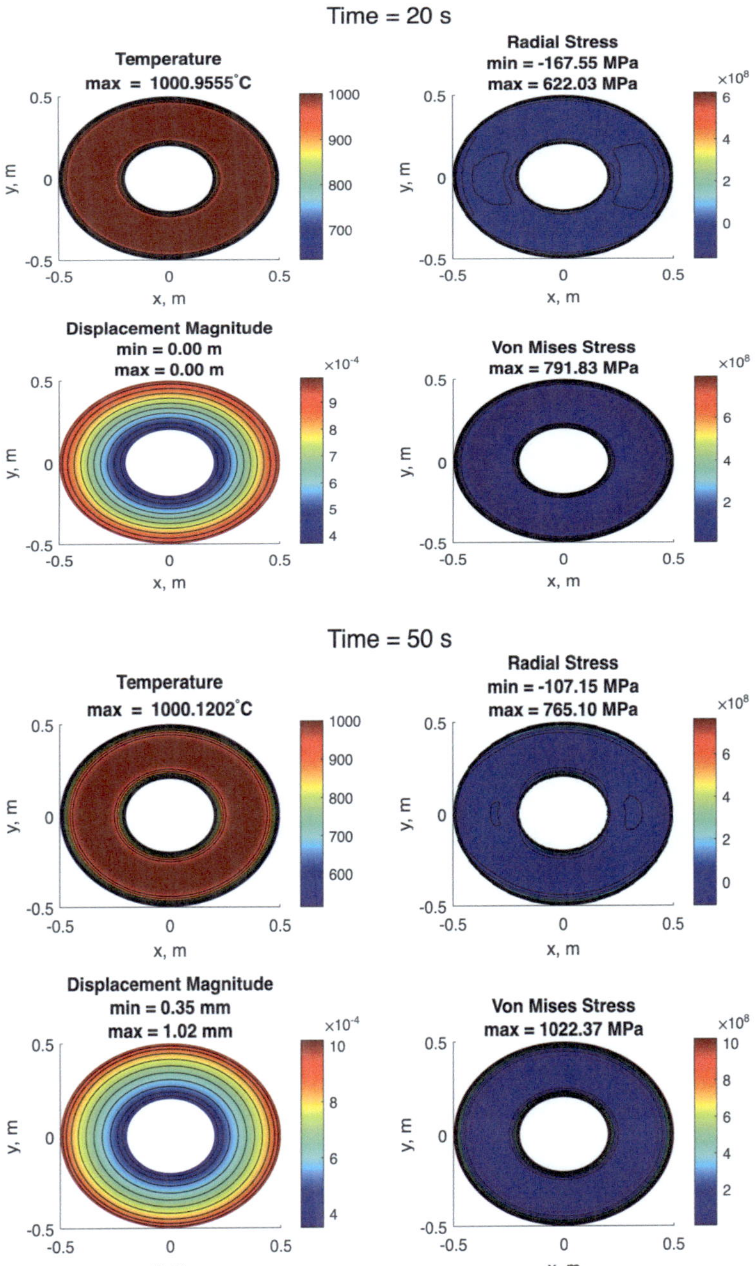

Fig. 5.4 (continued)

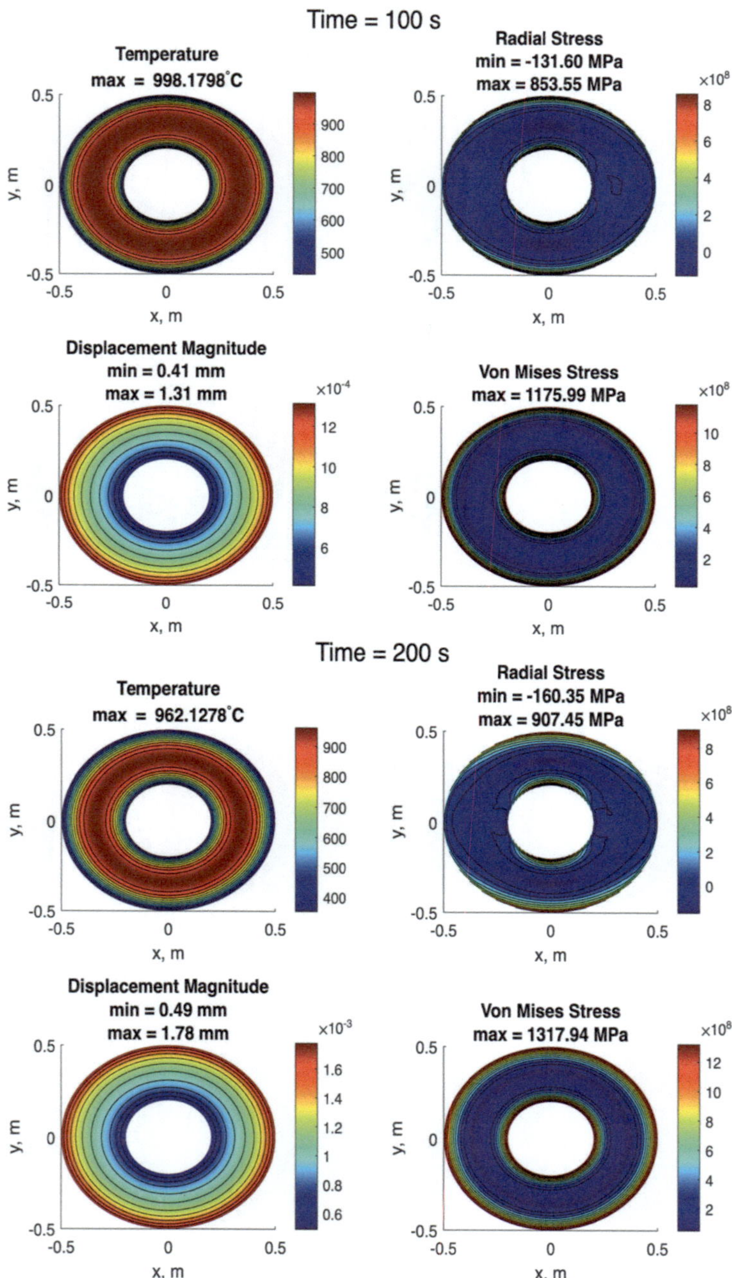

Fig. 5.4 (continued)

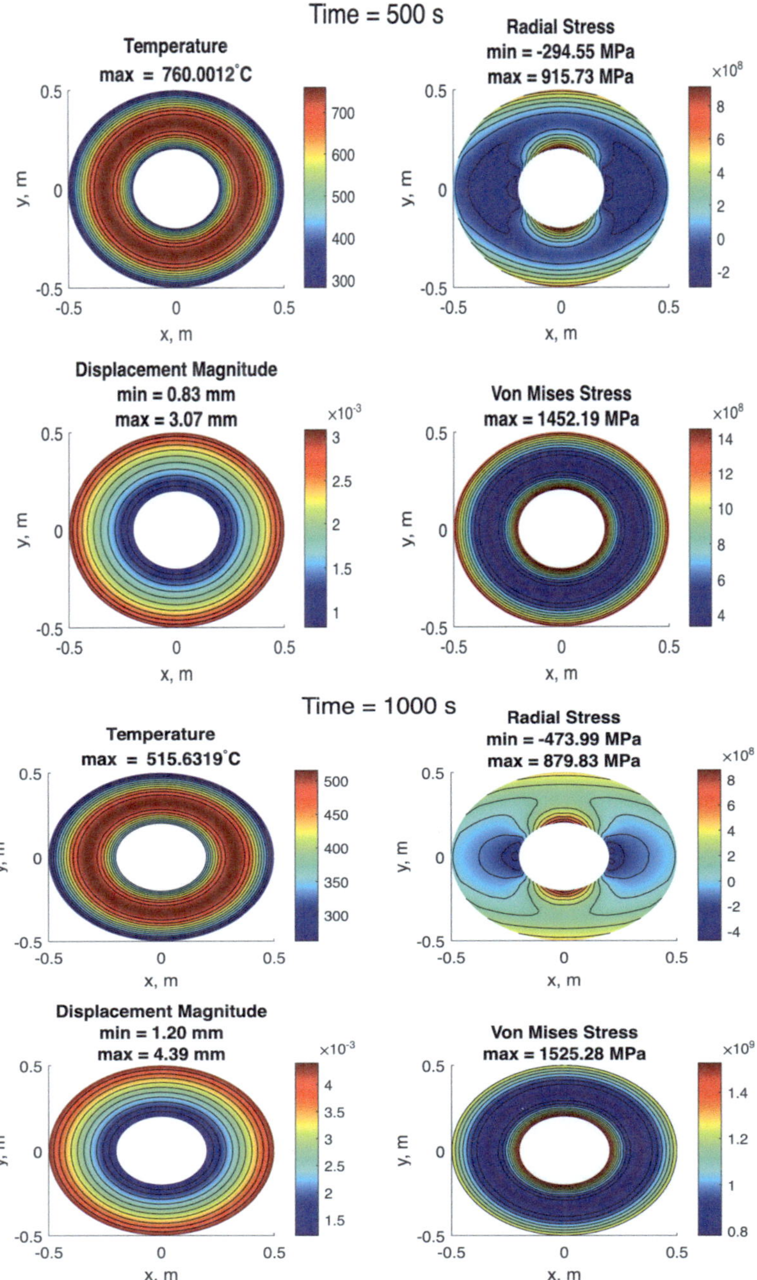

Fig. 5.4 (continued)

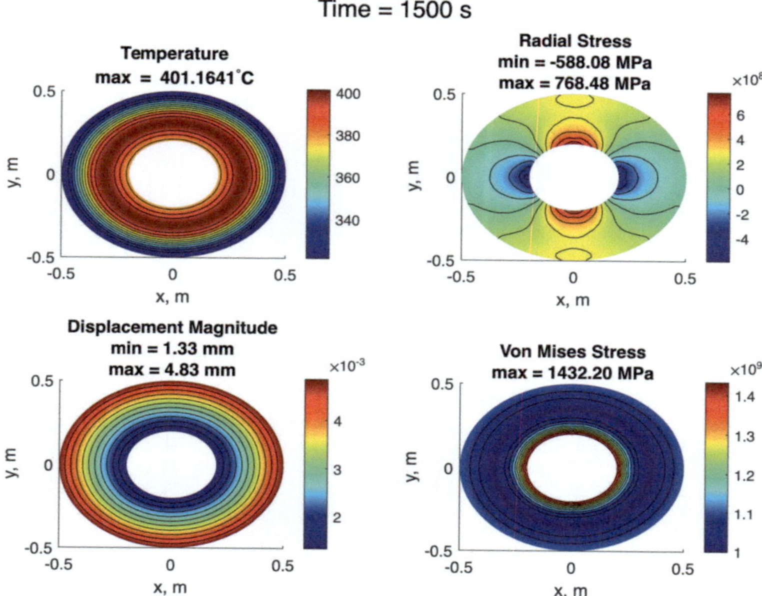

Fig. 5.4 (continued)

Fig. 5.5 Von Mises stress evolution with time along the radial distance

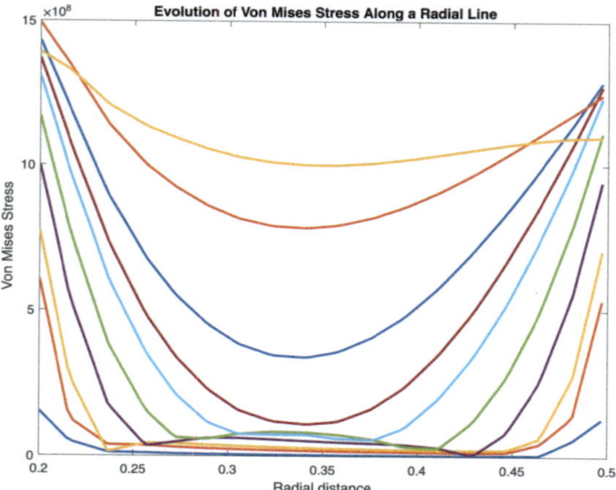

Surface Heat Transfer Coefficient Function

```matlab
function h = hFcn(location,state)% function to model a time dependent boundary
condition

h = 1500*(1 - state.time/1500.0);

end
```

Plot Results Function

This helper function plots the temperature distribution, radial stress, hoop stress, and von
Mises stress.

```matlab
function plotResults(model,R,modelT,Rt,tID)
figure
subplot(2,2,1)
pdeplot(modelT,'XYData',Rt.Temperature(:,tID), ...
                'ColorMap','jet','Contour','on')
title({'Temperature'; ...
      ['max = ' num2str(max(Rt.Temperature(:,tID))) '^{\circ}C']})
xlabel 'x, m'
ylabel 'y, m'



subplot(2,2,2)
pdeplot(model,'XYData',R.Stress.srr, ...
                'ColorMap','jet','Contour','on')
title({'Radial Stress'; ...
      ['min = ' num2str(min(R.Stress.srr)/1E6,'%3.2f') ' MPa']; ...
      ['max = ' num2str(max(R.Stress.srr)/1E6,'%3.2f') ' MPa']})
xlabel 'x, m'
ylabel 'y, m'

subplot(2,2,3)
pdeplot(model,'XYData',R.Displacement.Magnitude, ...
                'ColorMap','jet','Contour','on')
title({'Displacement Magnitude'; ...
      ['min = ' num2str(min(R.Displacement.Magnitude)*10^3,'%3.2f') ' mm']; ...
      ['max = ' num2str(max(R.Displacement.Magnitude)*10^3,'%3.2f') ' mm']})
xlabel 'x, m'
ylabel 'y, m'

subplot(2,2,4)
pdeplot(model,'XYData',R.VonMisesStress, ...
                'ColorMap','jet','Contour','on')
title({'Von Mises Stress'; ...
      ['max = ' num2str(max(R.VonMisesStress)/1E6,'%3.2f') ' MPa']})
xlabel 'x, m'
ylabel 'y, m'

sgtitle(['Time = ' num2str(Rt.SolutionTimes(tID)) ' s'])
end
```

5.2 Thermal Stresses in Welding

Background Information

Just like quenching, the process of welding is a unique multiphysics problem that involves heating from a traveling heat source, material addition, melting, fluid flow and mixing of species at high temperature, cool down and solidification and associated distortion and residual stresses including elastic-plastic behavior of the material. In this example a number of these problem features have been explored, namely heat input from a traveling heat source, heating and cooling, and elastic deformation and residual stress formation. To keep the complexity of the problem manageable, material addition, fluid flow and plastic behavior of material have not been included here.

Create Geometry of a Flat Plate for Thermal Analysis

Create the model for Thermal analysis. The geometry is that of a flat plate. Only half the plate is modeled and symmetry conditions will be used in the model.

```
modelT = createpde('thermal','transient');
```

Create the geometry

```
W = 0.2;% width (m)
L = 0.8;% length (m)
H = 0.05; % height/thickness (m)
gm = multicuboid(W,L,H); %create a rectanngular solid
gm =translate(gm,[-W/2,L/2,-H/2]); % move it so that the origin is at a corner
% the moving arc will be initiated here.
```

Assign the newly created geometry to the thermal model.

```
modelT.Geometry = gm;
```

Plot the geometry with the edge and face labels. We will use these geometric features to assign boundary conditions (Fig. 5.6).

```
figure;
pdegplot(modelT,'EdgeLabels','on','FaceLabels','on',"VertexLabels","on");
```

Fig. 5.6 Plate geometry

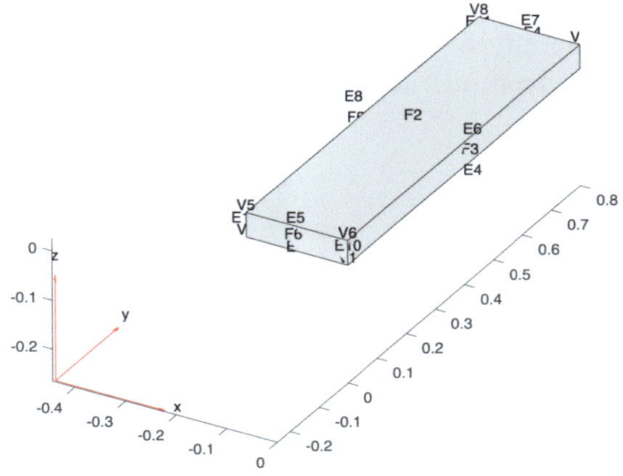

Fig. 5.7 Meshed geometry of
the plate

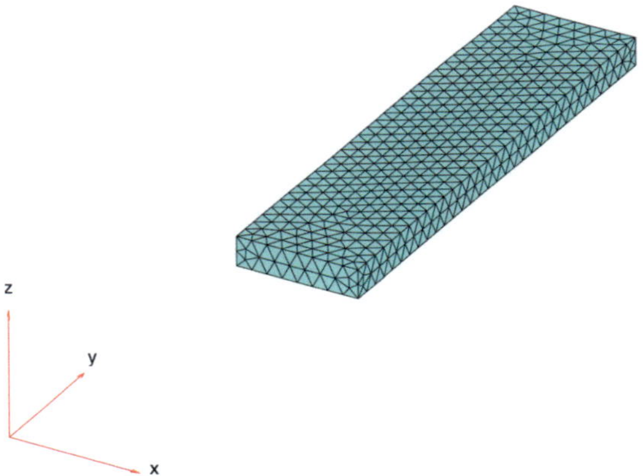

Generate a mesh. Use linear or quadratic geometric order for elements. An approximate
maximum size of the element is chosen here too (Fig. 5.7).

```
maxsize=0.025
```

maxsize = 0.0250

```
elemtype = "linear"
```

elemtype = "linear"

```
generateMesh(modelT,'Hmax',maxsize,'GeometricOrder',elemtype);

%draw the meshed geometry
figure
pdemesh(modelT);
```

Specify the thermal material properties of the disc necessary for transient analysis. The three properties needed in the analysis are specific heat, thermal conductivity and mass density.

```
alphad = 1.44E-5; % Diffusivity of disc
Kd = 51;% thermal conductivity,(W/mK)
rhod = 7100;% density, (Kg/m^3)
cpd = Kd/rhod/alphad; % Specific Heat,(J/kgK)
thermalProperties(modelT,'ThermalConductivity',Kd, ...
                        'MassDensity',rhod, ...
                        'SpecificHeat',cpd);
```

For the welding process the main thermal boundary condition is the heat flux input from a moving heat source, i.e. an arc. Researchers have modeled this moving heat source using different forms of a Gaussian function that is centered around an instantaneous location of the arc. This location changes as the arc moves along the welding path (Fig. 5.8).

$$q = \frac{6\sqrt{3}Q}{\pi\sqrt{\pi}\,\text{abc}} \exp\left(\frac{-3(x + v(\tau - t))^2}{a^2}\right) \exp\left(\frac{-3y^2}{b^2}\right) \exp\left(\frac{-3z^2}{c^2}\right)$$

where Q is the heat input from the arc in W, v is the speed of the arc movement and a, b and c are the semi-axes of the ellipsoid of influence of the arc in three dimensions.

For the rest of the exposed surfaces of the plate we use a convective heat transfer boundary condition with a specific value of h and an ambient temperature.

The heat input from the arc is modeled using the qFcn function on the top surface of the plate

```
thermalBC(modelT,'Face',[1:6],"ConvectionCoefficient",10,"AmbientTemperature",2
5);
% the top surface will
thermalBC(modelT,'Face',2,"HeatFlux",@qFcn);
```

Fig. 5.8 Schematic of the welding problem

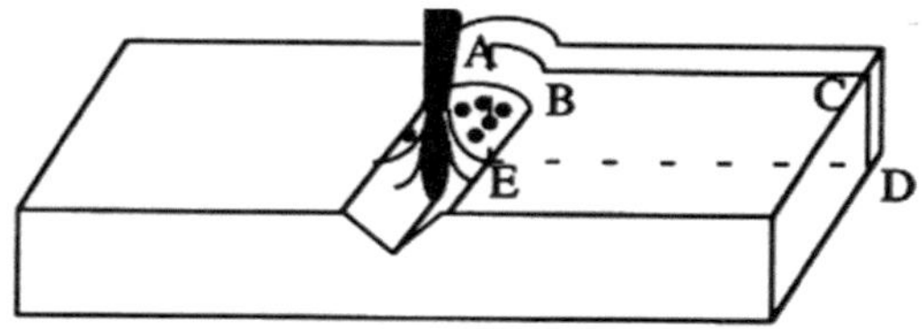

Schematic for a single-pass GMA weldment.

Set the initial temperature. The entire model is set at room temperature.

```
thermalIC(modelT,25);
```

Solve the model for the times used (Figs. 5.9 and 5.10).

```
tlist = [0:1:400];% the model is solved for 400 sec with an interval of 1 sec.
Rt = solve(modelT,tlist); % the solution process is carried out and stored in
Rt
 %Plot at a certain time step
 step = 82

step = 82

 temp = Rt.Temperature(:,step);
 pdeplot3D(modelT,'ColorMapData',temp);
```

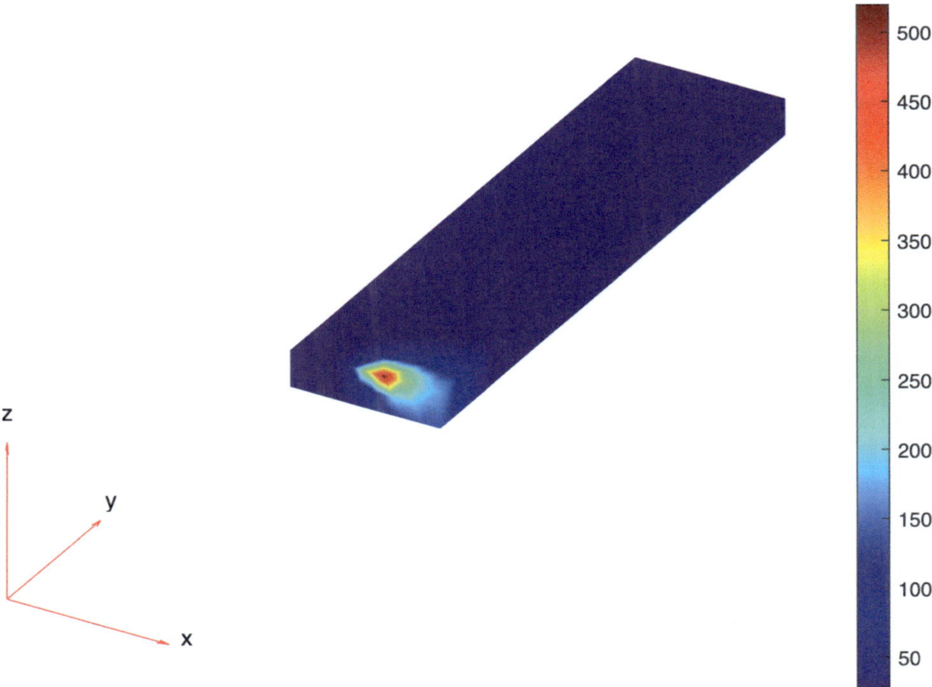

Fig. 5.9 Temperature profile at a certain instant

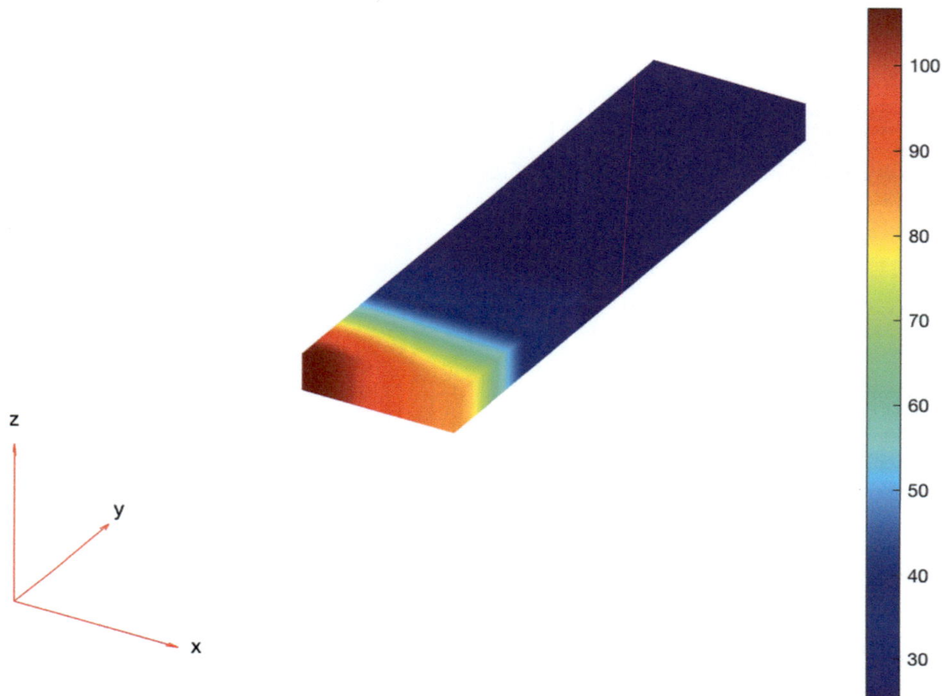

Fig. 5.10 Temperature profile at the end of analysis time

```
% Create a video of the traveling heat source
figure
m = 400;
for j = 1:m
    temp = Rt.Temperature(:,j);
    pdeplot3D(modelT,'ColorMapData',temp);
    M(j) = getframe;
end
```

Plot the temperature variation with time at three locations on the plate (Fig. 5.11).

Fig. 5.11 Temperature profile
at three different locations

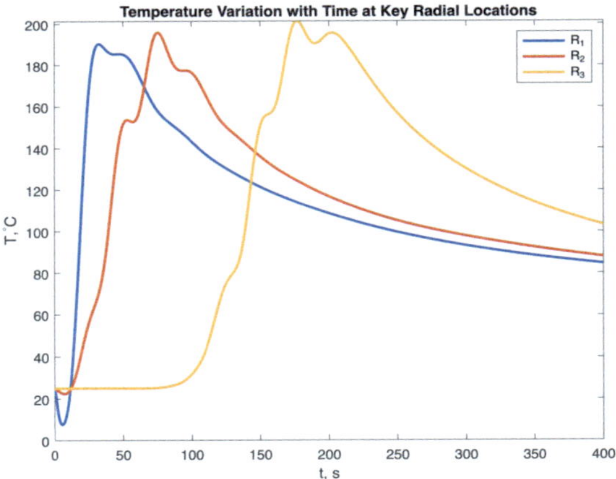


```
iTR1 = interpolateTemperature(Rt,[0;0;0],1:numel(Rt.SolutionTimes));
iTR2 = interpolateTemperature(Rt,[-0.05;0;0],1:numel(Rt.SolutionTimes));
iTR3 = interpolateTemperature(Rt,[-0.15;0;0],1:numel(Rt.SolutionTimes));
%
figure
plot(tlist,iTR1,'LineWidth',2)
hold on
plot(tlist,iTR2,'LineWidth',2)
plot(tlist,iTR3,'LineWidth',2)
title('Temperature Variation with Time at Key Radial Locations')
legend('R_1','R_2','R_3')
xlabel 't, s'
ylabel 'T,^{\circ}C'
```

Structural Analysis: Compute Thermal Stress

Create a 3D static structural analysis model.

```
model = createpde('structural',"static-solid");
```

Assign the geometry and mesh used for the thermal model.

```
model.Geometry = modelT.Geometry;
model.Mesh = modelT.Mesh;
```

Specify the structural properties of the disc.

```
E = 210E9; %Young's Modulus
nu = 0.29; %Poisson's Ratio
alpha = 1.2E-5; %coefficient of thermal expansion

structuralProperties(model,'YoungsModulus',E, ...
                          'PoissonsRatio',nu, ...
                          'CTE',alpha);
```

Constrain the model to prevent rigid motion. Apply the Symmetric Boundary condition on the face that is in the middle of the whole plate. Also, apply fixed boundary condition to one edge of that face.

```
structuralBC(model,'Face',6,"Constraint","symmetric");
structuralBC(model,'Edge',1,"Constraint","fixed");

%structuralBC(model,'Reference',[0,0],'Radius',20)
```

Specify the reference temperature that corresponds to the state of zero thermal stress of the model. We use the initial temperature of the plate as the reference temperature. Now when things are heated and cooled in a non-uniform fashion the stress build-up happens.

```
model.ReferenceTemperature = 25;
```

Specify the thermal load by using the transient thermal results Rt. The VonMises stress at the end of the analysis time and the displacement contours are plotted. For the definition of the plotResults function, see Plot Results Function. The results are shown in Figs. 5.12, 5.13 and 5.14.

```
for n = 2:numel(Rt.SolutionTimes)
structuralBodyLoad(model,'Temperature',Rt,'TimeStep',n);
R = solve(model);
end

pdeplot3D(model,'ColorMapData',R.VonMisesStress);
```

```
pdeplot3D(model,"ColorMapData",R.Displacement.Magnitude);
```

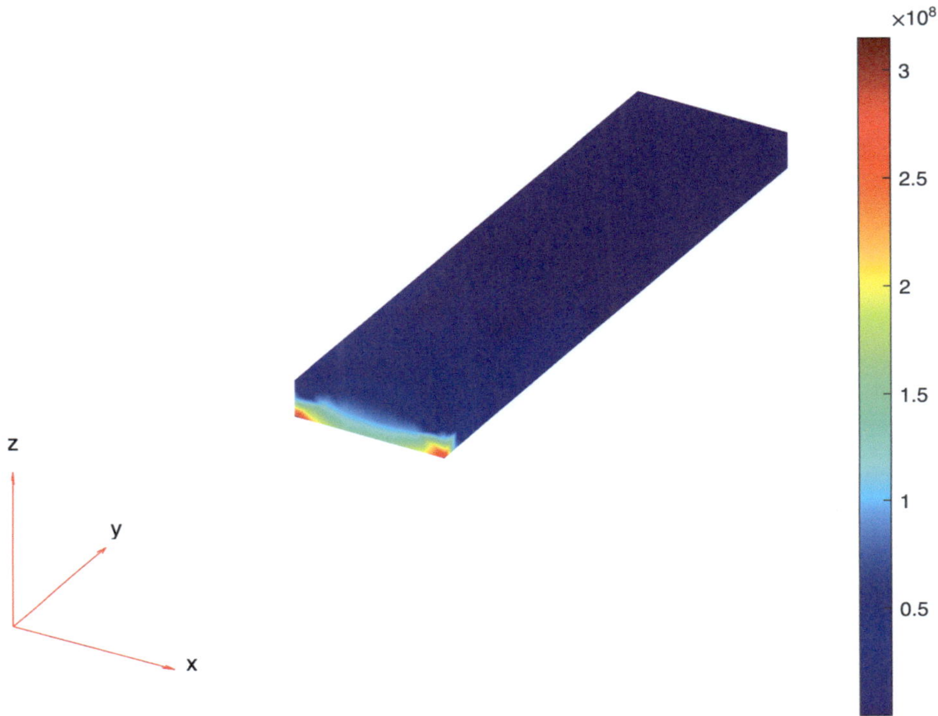

Fig. 5.12 Von Mises stress distribution

```
pdeplot3D(model,"ColorMapData",R.Displacement.z);
```

Surface Heat Transfer Coefficient Function

```
function q = qFcn(location,state)%function to plot the heat input from a moving
heat source
 a = 0.01;
 b = 0.01;
 c = 0.01;
 Q = 3000;
 speed = 0.001;
 q = (6*sqrt(3)*Q/(pi*sqrt(pi)*a*b*c))*exp(-
(location.x+speed*state.time)^2/(a^2))*exp(-(location.y)^2/b^2)*exp(-
(location.z)^2/c^2);
```

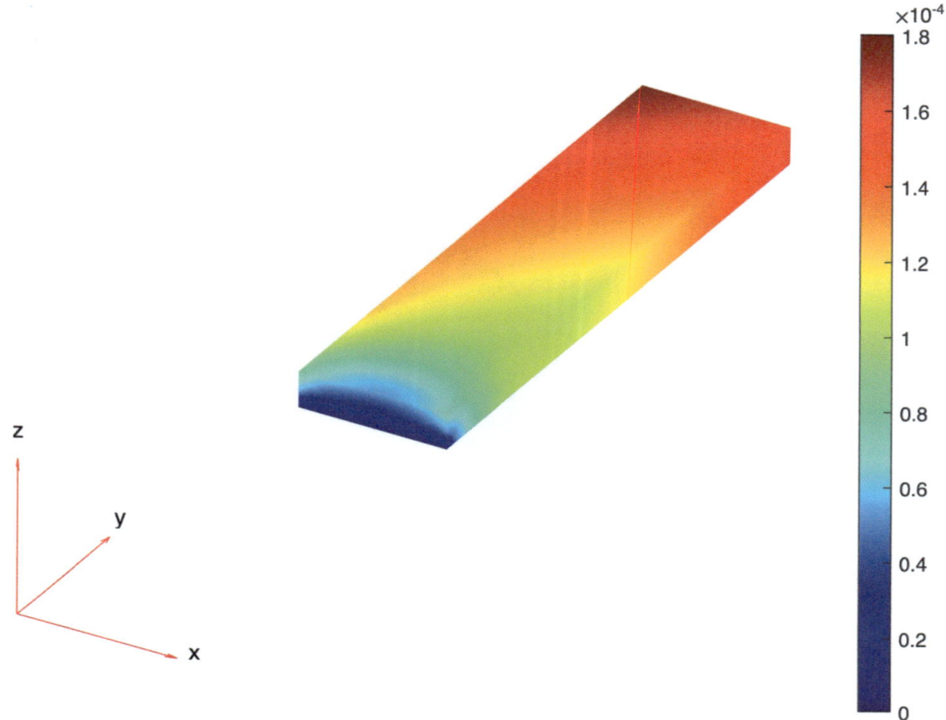

Fig. 5.13 Total displacement contours

5.3 Flux Density in an Electric Motor

Magnetic Flux and Field Calculation using the Electromagnetic module.

In this example we will find the static magnetic field induced by the stator windings in a four-pole electric motor. We approximate the motor using a 2-D model. The geometry consists of three regions:

Two ferromagnetic pieces: the stator and the rotor, made of transformer steel.

The air gap between the stator and the rotor.

The armature copper coil carrying the DC current.

The magnetic permeability of air and of copper are both close to the magnetic permeability of a vacuum, $\mu = \mu 0$. The magnetic permeability of the stator and the rotor is $\mu = 8000\ \mu 0$. The current density J is 0 everywhere except in the coil, where it is 1000 A/m^2.

The geometry of the problem makes the magnetic vector potential symmetric with respect to the y-axis and x-axis. Therefore, for the domain a quarter of the motor is drawn.

The magnetic potential on all the boundaries is assumed to be zero (Fig. 5.15).

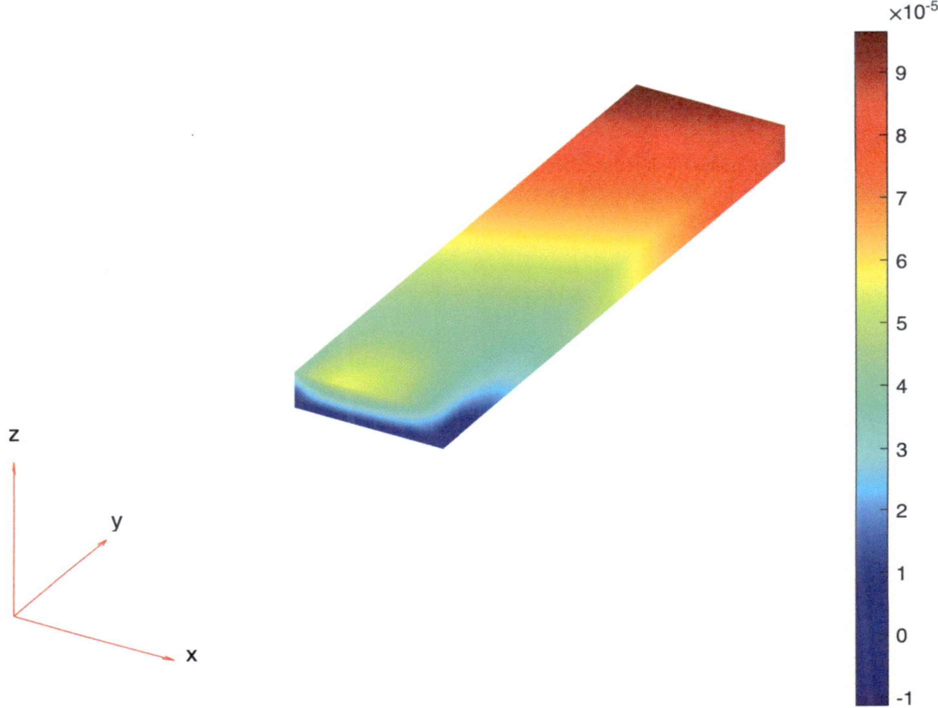

Fig. 5.14 Displacement contours in Z direction

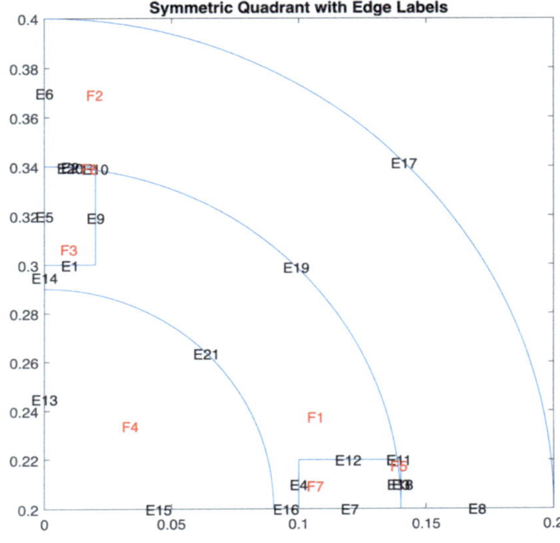

Fig. 5.15 The problem geometry, a quarter of the motor section with two poles

```
 model2D = createpde('electromagnetic','magnetostatic');% creating the model
definition
 radius = .20;%defining a base dimension
 R1 = [3 4 -radius 0 ...
       0 -radius 0 0 2*radius 2*radius]';
 R2 =[3 4 0 radius radius 0 0 0 radius radius]';
 R3 = [3 4 0 radius/10 radius/10 0 radius*3/2 radius*3/2 (radius*3/2+.04)
(radius*3/2+.04)]';
 R4 = [3 4 radius/2 (radius/2+.04) (radius/2+.04) radius/2 radius radius
11*radius/10 11*radius/10]';

 C1 = [1 0 radius radius 0 0 0 0 0 0]';
 C2 = [1 0 radius .14 0 0 0 0 0 0]';
 C3 = [1 0 radius .09 0 0 0 0 0 0]';
 gm = [R1 C1 R2 R3 R4 C2 C3];%Defining severaal different geometries to create
the deesired
 %geometry
 sf = '(C1+C2+C3)-R1-R2+R3+R4'; % the desired geometry is created
 ns = char('R1','C1','R2','R3','R4','C2','C3');
 g = decsg(gm,sf,ns');
 geometryFromEdges(model2D,g);
 %Plot the geometry displaying the edge labels annd face labels.

 figure
 pdegplot(model2D,'EdgeLabel','on','FaceLabel','on');
 axis equal
 title 'Symmetric Quadrant with Edge Labels';
```

Create an electromagnetic model for axisymmetric magnetostatic analysis and assign the geometry.

Specify the vacuum permeability value in the SI system of units.

```
model2D.VacuumPermeability = 1.2566370614E-6;
```

Specify a relative permeability of 1 for all domains.

```
electromagneticProperties(model2D,'RelativePermeability',1);
```

Now specify the large relative permeability of the core.

```
% the outer casing/stator and the rotor core have this high permeability
electromagneticProperties(model2D,'RelativePermeability',8000, ...
                          'Face',[2 4]);
```

Specify the current density in the coil.

```
electromagneticSource(model2D,'CurrentDensity',1000,'Face',[3 6 5 7]);
```

Assign zero magnetic potential on the outer edges of the air domain as the boundary condition.

```
electromagneticBC(model2D,'MagneticPotential',0,'Edge',[13 14 5 6 15 16 7 8])
```

ans =
  ElectromagneticBCAssignment with properties:

    MagneticPotential: 0
          RegionID: [13 14 5 6 15 16 7 8]
        RegionType: 'Edge'
        Vectorized: 'off'
           Voltage: []

Generate a mesh.

```
generateMesh(model2D,'Hmin',0.001,'Hgrad',2,'Hmax',.01);
pdemesh(model2D)
```

The meshed geometry is shown in Fig. 5.16. Solve the model

```
R = solve(model2D); %solve the problem
Bmag = sqrt(R.MagneticFluxDensity.Bx.^2 + ...
           R.MagneticFluxDensity.By.^2); %calculate the total flux density
from two components
pdeplot(model2D,"XYData",R.MagneticPotential,"Contour","on",'ColorMap',jet)
%plot magnetic potential contour


figure
pdeplot(model2D,"XYData",R.MagneticPotential, ...
               "FlowData",[R.MagneticField.Hx, ...
                          R.MagneticField.Hy], ...
               "Contour","on","FaceAlpha",0.5); %plot the
magnetic potential vectors
title ("Magnetic Potential and Field")
```

The Magnetic field potential and magnetic flux densities are shown in Figs. 5.17, 5.18 and 5.19.

```
pdeplot(model2D,'XYData',Bmag,'FlowData',[R.MagneticFluxDensity.Bx
R.MagneticFluxDensity.By],"Contour","on")% plot the magnetic flux density
vectors
```

Fig. 5.16 Meshed
geometry of the motor section

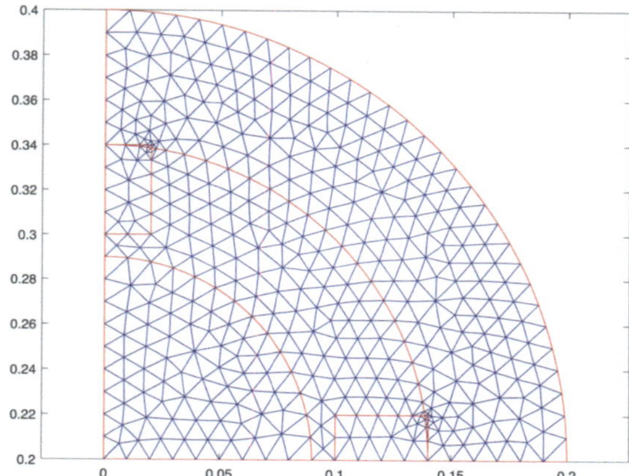

Fig. 5.17 Magnetic field
contour

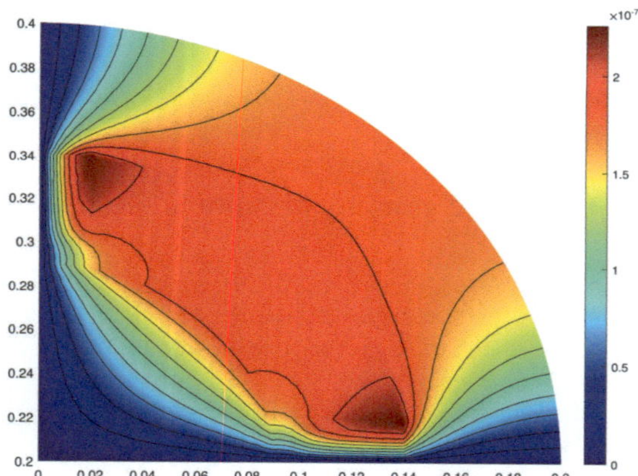

Fig. 5.18 Magnetic field
vectors

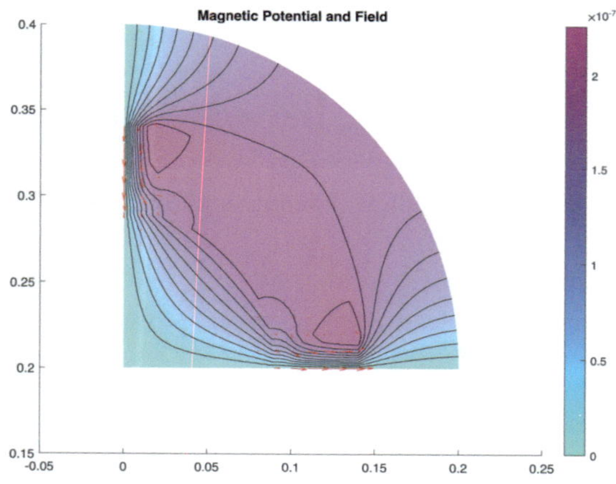

Fig. 5.19 Magnetic flux density contours and vectors

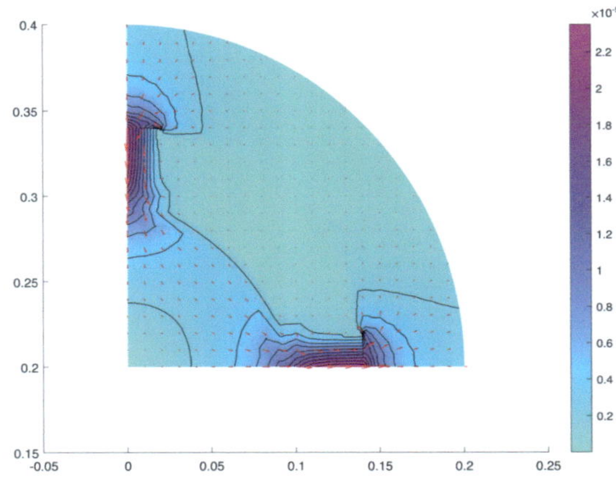

5.4 Summary

In this chapter we presented three examples showcasing some of the additional features of the PDE toolbox. The first two examples deal with thermal stress induced during a heat treatment process and a welding process. Users should be aware that these two problems were solved as linear elastic problems even though in reality, these types of problems could involve plastic (non-linear) behavior of materials. The third example had to do with calculating magnetic field in an electric motor due to the coils of wire in the stator.

FSC
www.fsc.org
MIX
Papier aus verantwortungsvollen Quellen
Paper from responsible sources
FSC® C105338